IMPROVING THE FOXHOLE: WARD'S WAY

AF722076

IMPROVING THE FOXHOLE: WARD'S WAY

REFLECTIONS ON LEADERSHIP, LIFE, AND SERVICE

WILLIAM E. "KIP" WARD

Curry Brothers Publishing

Improving the Foxhole: Ward's Way

Copyright © 2026 by William Ward
Produced in association with Vibranium Media Group

The scanning, uploading, and distribution of this book without permission is a theft of the author's intellectual property. No part of this publication may be reproduced, stored in a retrieval system, or transmitted in any form or by any means, electronic, mechanical, photocopy, recording, or any other, except for brief quotations in printed reviews, without the prior permission of the publisher. If you would like to request permission for use of material from the book or any additional information, please contact Curry Brothers Publishing. Thank you for your support of the authors' rights.

Curry Brothers Marketing and Publishing Group
P.O. Box 247
Haymarket, VA 20168
currybrotherspublishing.com

First Edition: April 2026

All rights reserved.

Cover Design and Graphics: Vibranium Media Group
Copy Editing: JS Johnson Company
(Chantel Johnson & Tiffany Johnson)
Executive Editing: Joniece Jackson

ISBN: 978-1-966480-16-7 Hard
ISBN: 978-1-966480-44-0 Soft
ISBN: 978-1-966480-19-8 eBook

To my sweetheart Joyce, the love of my life and my lifetime partner...

and

To our two amazing children, KJ and Kahni who make us proud everyday.
This journey is yours also.

ABOUT THE TITLE

My appreciation for "improving the foxhole" began on Fort Bragg, North Carolina, in my early days as the platoon leader for 37 soldiers: the 3rd Platoon of A Company, 3rd Battalion (Airborne), 325th Infantry in the 2nd Brigade of the 82nd Airborne Division.

At the conclusion of a routine field training exercise (FTX), where my company had been in the field for about 11 days training for combat, I received a call from my company commander that we had concluded the FTX. He indicated the company's First Sergeant, the top enlisted soldier in the company, would provide redeployment instructions to my platoon sergeant (PSG), the top enlisted soldier of the platoon. My PSG and I shared a fighting position, aptly called foxholes in those days. Foxholes were prepared positions used to protect soldiers from enemy fire and for fighting. The better the foxhole, the greater the likelihood that it would keep you safe. Therefore, you were always looking for ways to improve it.

When my company commander told me the FTX was over, I immediately told my PSG, took off my steel pot (helmet) and laid my rifle down on the rim of our foxhole. My PSG stopped me in my tracks as I was "chilling" saying, "What are you doing, Lieutenant? As long as we occupy this foxhole, we work to improve it. This will save our lives." That was after a training exercise and we were not at war, but the lesson stuck: Always work to improve the foxhole as long as you're occupying it.

That lesson has been with me since those early days of my career. Every position I've held since then has been my foxhole, whether a command or staff position. I've worked to improve it as long as I occupied it. I have imparted that discipline to everyone in the organizations I've led as well, even into my civilian capacities.

From time to time, you need to get out of your foxhole and look at it from the perspective of others. Be observant. Be quick to listen and listen well. You will see gaps and ways to improve and enhance your environment and that of others. Foxhole lessons are life lessons. "Ward's Way" is the result of how I tried to improve one of the foxholes I found myself in.

During my tenure as the Commander of the NATO Stabilization Force in Sarajevo, Bosnia-Herzegovina, my headquarters was located at Camp Butmir. In addition to the military forces from myriad NATO and NATO friendly countries occupying living and workspaces on Camp Butmir, local residents of Sarajevo were also employed on Camp Butmir in varying capacities. A few months into my one-year command tour, I had traveled overnight to an outlying installation on a command visit to one of the military sectors as I often did. Upon returning to Camp Butmir early in the morning I saw the local citizens from Sarajevo getting off a bus some distance away from the Camp entrance and walking to the gate.

What I saw was not acceptable. The people had to walk through a muddy path in order to get to the gate. When I found out that was their normal and most direct route and had been taking that route for years, I said to myself, that needs to be corrected. Once in my office I asked the camp supervisor and public works director to come see me. My question was, is there a reason that the pathway used by the civilian workforce couldn't be paved so the people didn't have to walk and ride their bicycles through mud. There was none, other than no one had ever paid any attention to it and asked about it. Given standard bureaucratic procedures, their funding and construction request was eventually submitted but as my command tour was winding down, the pathway had still not been paved. I then made a serious point about getting the paving done and I told the people that I would do my best to push the project through prior to my departure. It still hadn't happened when I changed command.

Not long after I left Sarajevo, the European Command Chief of Staff, Maj Gen Bill Catto, visited there and came back with a photo of the now paved thoroughfare from the bus stop to the Camp Butmir gate. Signs with the name, Ward's Way, were posted at both ends of the bike path and walkway. The local civilian workforce had requested the pavement be named for me, remembering I had championed the cause.

Taking care of my people has always been a priority, as stated in my 10th Mountain Division Commando Brigade motto - "People First, Mission Always!"

"Improving The Foxhole: Ward's Way" is a reflection of that ever-present priority.

CONTENTS

FOREWORD

By Ambassador Linda Thomas-Greenfield

I am not sure when I met General Kip Ward. I am guessing it was sometime in 2006 when I was appointed Principal Deputy Assistant Secretary of State for Africa, and he was the Deputy Commander at EUCOM. I was tasked along with my DOD colleagues to travel around the continent of Africa to cultivate/introduce the concept of a new unified Combatant Command for the continent of Africa.

The continent of 54 countries was managed by three different commands, three different generals and different approaches. Thus, changing this structure was no easy task as there were many different interests particularly on the continent and within the military. But for my colleagues in the Africa Bureau and across the continent, the initiative represented a new interest in the continent and a recognition of the importance of the military in that three-legged stool of diplomacy, development, and defense. AFRICOM had to also manage relations with the Middle East Bureau which had responsibility for the four North African countries in that bureau.

General Ward was immediately identified as the person for the job with the support of Secretary of State Condoleezza Rice, who knew him personally. Kip who? I remember thinking. And then I met him. General Ward was a military guy. He lived the part. He looked the part. He walked the part, and he spoke the part. When he entered the room, the room came to attention. He took on the role with purpose and commitment. Built the command in real time or as the military like to say, "Built the plane while flying it!" He was the right person for the job.

In 2008, I was appointed Ambassador to Liberia, and found that AFRICOM was already operational and effective on the ground in this country that was coming out of a decade-long civil war. Liberia's army had been shattered by the war. AFRICOM was tasked to recruit, train and mentor the first 200 members of the new Liberia Army. General Ward would be my first visitor to the embassy, days after my arrival, to visit his troops and see their work. As I escorted General Ward from his plane, I could see the recruits stand taller, salute more exact, and march with more purpose. Even I stood erect!

In our meeting with President Ellen Johnson Sirleaf, General Ward exhibited not only his military skills but his diplomatic acumen. He shared with Sirleaf his assessment of her budding military and his recommendations on what her future security needs were. We discussed the need for a guard to patrol the coast of Liberia where illegal fishing was stealing the country's resources. He committed to assisting her building the coast guard and delivering coast guard patrol boats to them once they were trained. It was a proud moment the day the boats were delivered, and General Ward would escort the president and me for an initial trip in Liberia's coastal waters.

General Ward's trip to Liberia was enhanced by a secret weapon—his wife Joyce, who in her own right would make a name for herself in Liberia (and other countries in Africa), meeting with the communities served by the military and Liberian women who were victims but also warriors in the fight for peace.

The impact General Ward had on the entire continent of Africa was immense. He was not only respected by his troops, but military commanders and leaders across the continent. He was a tool of U.S. soft power but a powerful voice for our military might and capabilities.

But there is more to this general than his impact on Africa. He was a role model and mentor to many generations of U.S. military men and women, to young people graduating from HBCUs and other colleges and universities, to young people growing up in working class families. He was a role model and a partner for me during the last years of my career.

In his retirement, he continues to have impact and command

respect. I am proud to have known him not just as a colleague but a friend. "Kip who?" is now my friend, General Kip Ward.

His lessons on leadership, his life experiences and his service to our country during 9/11 at the Pentagon, in Bosnia-Herzegovina, Egypt and Israel among other tours around the world will be a guide for the next generations.

Improving The Foxhole: Ward's Way is a must read!

Linda Thomas-Greenfield
Ambassador (Retired)
Former U.S. Ambassador to Liberia and the United Nations

ENDORSEMENT

By Lieutenant Colonel (retired) Bruce Moody

General William "Kip" Ward modeled an uncharacteristic leadership style from most leaders in his generation and throughout history and it made all the difference.

He possessed a rare gift for connecting with everyone on their level, making them feel valued, understood, and respected. One striking example of this was his retirement ceremony. The line of well-wishers eagerly waiting to congratulate him stretched for nearly an hour, yet everyone waited patiently for their turn to speak with him. It was so striking that other officers and their wives commented on the unusual sight of so many well-wishers patiently waiting to bid farewell.

Soldiers, sailors, airmen, marines and everyday people all stood in line to express their gratitude. In that moment, I realized that no other general in the Army commanded such unwavering loyalty. His influence extended across all ranks and military services. He was cherished and respected by the service secretaries and other general officers, but he was equally admired by the Pentagon security force, the custodial staff, and the cafeteria workers. People from all walks of life, from every rank, profession, and persuasion, trusted him as one of their own. They were willing to wait an hour because they shared personal stories and cherished memories of him, and they were determined to remind him of his significance in their lives.

As a Brigadier General, at Fort Bragg, he mentored a cohort of officers who went on to become senior military leaders for the next three generations. This single cohort produced more general officers than any leader of his time. These officers advanced to lead at the highest levels of the Department of Defense. His leadership style was grounded in a humanistic approach. He ensured that his behavior aligned with his values. By modeling infectious enthusiasm, moral courage, presence, and accessibility, he empowered other leaders to act courageously. His example served as a clear message that leaders were empowered and expected to lead with moral courage and that they were accountable for their decisions. This gave them permission to operate fairly and consistently, even in the face of organizational pressure to conform and fit in.

As Deputy Director for Operations (DJ-3) in the Pentagon, he created an environment where everyone could thrive. Watching him interact with people was like witnessing "everything, everywhere, all at once." It was not uncommon to see him asking a young airman where they were from or stopping to speak to a junior staffer about their goals and aspirations while headed to meet with senior officials or respond to a national emergency. No matter the situation, the common refrain was, "Hey Buddy!," and it was so heartfelt that each recipient felt like the only person in the room. He had the rare ability to slow down and be fully present for those he encountered without diminishing the urgency or the importance of what lay ahead.

In today's AI-driven world, where real-time innovation and problem-solving are commoditized and reduced to algorithms, humanistic leadership is vital to empower individual team members.

General Ward managed to lead with humanity. This people-centered leadership style starts with self-awareness and a deliberate act to live authentically and demonstrate unapologetic kindness and caring, even in the face of his remarkable accomplishments or seemingly insurmountable challenges. The antidote to his elevation is to engage others with genuine appreciation. These leadership lessons are equally applicable to generals and janitors, presidents and privates, captains of industry and captains of Infantry. All can benefit from the lessons herein.

Dear General Ward:

Sir, please permit me to write a short letter of support and enclose it with my holiday card. My cadets and I were privileged to assist you at your home during official holiday receptions. In addition, my cadets and I were honored to support many community events where you were the ranking officer. We were always moved and touched by your kindness, thoughtfulness, and generosity. Sir, I know of no individuals in the Stuttgart community that were more loved and idolized than you and your family!!

Perhaps Sir, you will recall that I was the JROTC instructor at Patch High School. I wore the uniform of our Army in some capacity, for 65 years—30 years activity duty, 4 years reserve, and 31 years as the JROTC instructor at the high school.

During my years in uniform, I had contact (both officially and unofficially) with countless senior officers of all our military services. Again, I can recall no individual more loved than you, Sir. Every individual who had any contact with you and your family had nothing but respect, admiration, and affection for you and your family.

It saddens all of us to read of recent events and we hope and pray that the harsh rulings will be reconsidered. Yours has been a long and distinguished career of dedicated service to our nation. Those of us who know you and loved you number in the thousands, probably in the 10s of thousands. Nothing can erode that love and respect. Absolutely nothing!!

As we approach the holiday period, you can be assured that the Ward family is in all our hearts, thoughts, and prayers. We pray that God will look over and protect you and yours and we wish you every possible happiness in your well-deserved retirement. May God continue to bless and look over your family!!

With kindest regards,
Very respectfully,
Jack Goldman, CW4, USA (Retired)
Stuttgart, Germany

PS. Always remember, Sir, you have an "army of supporters" who love you and stand behind you 100%!! That will never change! May God bless! Jack.

INTRODUCTION

I've lived a life that has been rich and full. I've been blessed to be married to a beautiful woman, the love of my life, for over 55 years; for that I'm very thankful. We have two wonderful children, their partners, and two amazing grandchildren for which I'm also thankful.

This autobiographical memoir is important for a few special reasons: 1) for my family to know how much I appreciate them, 2) for the people I've worked with over the years to know how I have valued our relationships, and 3) to document the ordinary and extraordinary experiences that have been meaningful to me and to others. I've just done what I've been asked to do to the best of my ability. In some small way these experiences may prove useful in life and leadership to someone else.

Over the years, when I've talked about my career experiences, adventures, and assignments, often the response is, "Wow! Those are such rich experiences. When are you writing your book?" So now it's my time to share what I've done and how I've done it. Now, a lot of what I've accomplished happened because of extraordinary people—family, colleagues, and friends—placed in my path. I've continued to learn from and reinforce lessons and actions I've picked up along the way. I'm so grateful for my early childhood years, the example of my parents and grandparents; I treasure the experiences of going to college and beginning my military career. I value the opportunity to put this down in my own words so those who are close to me and know me have a record that might be used to enhance the experiences and understandings of others from various spectrums. Plus, it's a relief to have them written, rather than just in my mind.

I hope others will read and understand the life of a boy from Baltimore with aspirations, not grand ones, but of seeing himself living out his purpose in a different place. A boy blessed with encouragement from parents, relatives and supportive people throughout his life. These people enabled me to "be the very best that I could be." I was taught to appreciate and take advantage of the things that the good Lord put in front of me and to do my best to provide others with those same opportunities.

To this day, I still have friends and associates from my time as a professor at West Point. We are so appreciative of those times. It's another case of developing ties that are like family that is beyond your blood family. From West Point we were fortunate enough to be selected to attend the United States Army Command and General Staff College at Fort Leavenworth, Kansas, and that was our next duty assignment.

Having served all over the world and with people from all backgrounds, something distinctive about an unfamiliar place always arises. People talk about their best tours or worst tours and assignments. We've lived in some unique places, across the United States and from Alaska to Hawai'i, in Europe, Korea, Africa and the Middle East. Joyce says, "The tour is what you make it. The tour is about the people, the things you do and the activities you engage in, not the location; that only provides the venue. The place is what you make it."

My retirement ceremony was a culmination of a career of service of which I am very proud. That ceremony was such a memorable reflection of a career all wrapped up in a formal review on Fort Myer with all elements of the Old Guard participating. Present on that sunny day were family and friends, people from my childhood, my classmates at Morgan State College that I graduated and served with; friends Joyce and I met over the years were in attendance from across the United States and overseas, all helping Joyce and me celebrate the occasion.

That is the essence of my story, and I'm just so blessed to have been able to do what I have done over the years. It is not that what I've done was so special, but each step along the way was significant and essential in setting the stage for what was ahead. I didn't always realize that at the time. I am thankful and satisfied, however, that I came to appreciate every experience with that knowledge. I don't envy others, and I don't covet anything. I've been blessed with my family, military family, and friends from all around the world. The good Lord has graced us and for that I'm thankful.

CHAPTER 1

EARLY DAYS to HBCU to ROTC

Kip—a name that is not on my birth certificate. William Edward Ward is the name on that and other legal documents. Yet, I can't remember when I wasn't called Kip. Around the time I was born, my mom's baby sister was reading a comic book series called "Kipsy Twins" I believe—it was 1949. She saw me—not her first but probably her closest and longest encounter with a real live baby—and after the introduction, called me Kip. Even when she talked about me, she would say Kip. Something about that comic book character gave her the idea to call me by that name, and it just stuck. Other than for legal documents, I've never used William, and I have never gone by Bill, Billy, or Willy. With family, friends, and professionally, it's always been Kip.

My boyhood years happened in the 1950s, growing up in Baltimore, Maryland, guided and nurtured by a World War II veteran father, Isiah Richard Ward, Jr. from Bedford County Virginia and a wise mother, Phyllis Mary Cashen, a Baltimore high school graduate. Although not college-educated, my parents were resourceful in many ways. With my younger sister, Christina Larnell Ward, we were a family of four.

We lived in a small Black community, Cross Keys Village, in northwest Baltimore on Falls Road. Although that small, tight-knit working class community no longer exists, it became a solid foundation for my existence and provided my first experiences of a village raising a child.

My first memories are of living in a small wooden frame house my parents rented and hearing my father say the house wouldn't stay warm. Yes, the house was in ill repair. In a way, that imperfection served as a catalyst for my parents to build their own home as they were tired of paying rent to a landlord who didn't keep the home in a decent living condition. Instead of living in someone else's frame house, my father chose to build his own home—and over the period of six years, he built a three-bedroom brick house in nearby Baltimore County.

My mother and other people shortened Dad's first name Isiah (intentional spelling) to "Ike." He didn't mind it, and like most nicknames, it seemed to fit his skills. Ike Ward was the brick mason, plumber, electrician, roofer, framer and finisher that the community and my family depended on.

It took six years for him and my mother to "pay as you go" in building the house, progressing as they were able to buy materials, with my mother, sister, and me assisting as the primary labor. I was a young boy with a front row seat to commitment, dedication, sacrifice, perseverance, and love, as I mixed cement, pushed a wheelbarrow, carried lumber, hauled bricks and cinder blocks, and more.

We used bricks that were retrieved from buildings that were torn down in the city of Baltimore to make way for other facilities and structures. We'd collect the used bricks, scrape the mortar off, clean them, and use those bricks in the construction of our home. It was and still is a solid brick home with concrete block walls. Today, although the interior has been renovated, it's in good shape and standing because of my father's dedication to doing a job the right way—something that I observed and imitated. From his example, I learned and understood the importance of having a goal with a strong and committed work ethic, and what it meant to be a part of a community of folks that cared.

For a few years, as our house was being built, we stayed with my maternal grandparents, Rev. Joseph and Marie Cashen; they lived nearby. Their house was the parsonage belonging to the church where my grandfather pastored. It, too, was a frame house with three small bedrooms upstairs, a "front room" as the living room was called, a dining room, a kitchen, and one small bathroom. It was fine; we made do. My grandparents had the front bedroom, and my sister and I had the middle bedroom. Access to the bathroom was through the back bedroom where my parents slept. We didn't know what we didn't have, and this worked.

During most of my young life I was around folks like my parents: blue collar people who worked hard every day. Many of the families worked in the factories, steel mills and shipyards of Baltimore. They had clarity and focus that their children's lives would be better than theirs and did everything they could to make that a reality. From the standpoint of seeking higher education, my generation stood firmly on their shoulders. The work that these women and men did for their families was honorable and that example continues even today.

My playmates were kids in my neighborhood. I remember building skateboard scooters. We used milk crates and put used roller skates on these 2x4's and attached the milk crates to them. Those were the skateboards of our time.

My first bicycle was a three-speed English racer. I must have been nine or ten years old. It probably took my parents a year to save up to buy it, but that was my Christmas present, and I was so happy. I would clean the rims and fly streamers from them. I added a horn and a generator to power small front and back lights. I was always adding to it and I kept that bicycle clean. I kept that bike for a good 30 years, but during one of my military moves, I finally got rid of it. To this day I wish I had kcpt it. I realized later how much it meant to me. I took pride in caring for that bike. At the time I didn't really appreciate how memories, reflections, and collector's items work. It helped me never to take things (or opportunities) for granted, someone sacrificed and worked hard to provide them.

That's a function of having something that you appreciate and just not taking it for granted or abusing it. The significance of being given something at that age was not lost on me. I knew it was a sacrifice for my folks, so that bike was meaningful and had value, and I took care of it.

I grew up going to my grandfather's church. My maternal grandfather, Rev. Joseph Henry Cashen, was the minister of an AME church in a section of Baltimore then called Cross Keys Village, which was later torn down to build a high school complex, Poly-Western High School. Rev. Cashen was known for his compassion and well-loved by his family, the community and his congregation.

Learning that my grandfather had also practiced boxing intrigued me and belied his pleasant demeanor. As he preached, my young, wandering mind often pictured him in a boxing ring. There were other stories of my grandfather as a sprinter. I was a pretty fast runner – a sprinter in high school – and will never forget my pre-teen races against him and how impressed I was to be beaten by him.

Ike Ward's Path

Having been born, raised and drafted in segregated Bedford, Virginia, and following his honorable discharge from the Army after World War II, my father found himself in Baltimore. Ike Ward worked mostly as a handyman/jack of all trades, as a townhouse apartment complex manager in Baltimore (work that included Saturday and Sunday morning trash pickup). Upon arriving in Baltimore, he worked closely with his dad, and both excelled with similar skills of working with their hands.

My father took advantage of his GI Bill and enrolled in various trade schools. Having been a construction combat engineer during World War II, his plumbing and electrical skills were advanced. He excelled especially as a cement and brick mason. In his spare time—evenings and weekends—he continued his journey of building our house. My mother, Phyllis Cashen Ward, worked as an office assistant at a local chemical factory in Baltimore and later as a file clerk.

As all of this was going on, I absorbed a lesson that stayed with me: how my father treated other people. He worked hard on his regular job, all day, every day and in the evenings and on weekends—all the while, he was building the house that his family would eventually move into. And after that was done, people would call him to do various home projects. If there was a plumbing problem or electrical problem, Dad was there. Oftentimes, the requests were for masonry needs.

Dad was an expert at laying brick and pouring concrete. It seemed like he never took time off. When he wasn't working on our home, he was working on someone else's home. Many times, on weekends and Saturdays I would be with him. I watched him never say no as people asked for help. I could see how tired he was, but his response would be yes, and he would show up. To this day that has great meaning for me; it's possibly why I have a hard time saying no and I don't mind being on that side of that equation.

From time to time my father borrowed machinery because of the relationships that he had established. He knew people who would lend him equipment, perhaps a bulldozer or backhoe to do some of the heavy digging. But most of the work was done by pick, shovel, and wheelbarrow, including digging out the basement. We hand mixed and poured cement for the footing and laid the blocks for the foundation.

As a teenager, I had no specific professional objective or goal for my life, but I had a vision. I liked anything automotive and had a fascination with trucks, cars, and tractors - almost any kind of vehicle. To this day I still have an affinity for those sorts of toys. I've always respected and been moved by the construction, assembling, and the building; being around and watching diverse people do what they do always inspires me.

I certainly enjoyed that growing up. Dad was a master with many skills, he even built a go-kart and a small tractor. He literally took used Briggs-Stratton lawn mower engines, welded tubular iron and metal into frames, mounted wheels, etc., and made these machines that I rode, zipping up and down our neighborhood. What we couldn't afford, Dad made. He was widely known for building the carnival rides his Shriners organization used for their programs in supporting children living with disabilities.

His skills were on display into my adulthood, as he made a U-Haul-type trailer for Joyce and me to use on our moves. Dad told us, "Instead of you renting these U-Hauls, I'm going to build you one!" Living up to my "my-Dad-can-do-anything" image, he made a trailer that served us well; so that every time we moved, we wouldn't have to rent one.

He had built small utility trailers for his gardening and handyman-type equipment. He didn't have a pickup truck, so—he built himself a trailer to hitch behind the family car! He'd previously built one of these trailers for the owners of the apartment complex he managed. Part of his job included driving one of their cars with the trailer attached as they moved annually from Baltimore down to Miami Beach each winter.

Life with Grandparents

Sunday dinners were a tradition, a constant part of growing up. After church, we gathered at my maternal grandparents' home. Their home was the parsonage, next door to the church, so the value of ritual was always present and convenient. My mother's three sisters, all my cousins, aunts and uncles would be there. Those Sunday gatherings stopped when the community was torn down to build a new high school complex in North Baltimore.

My early work life revolved around being with my dad and spending time with my grandfathers— Richard Ward, Sr. and Rev. Joseph Henry Cashen. I was blessed because I had the wonderful experience of growing up with both maternal and paternal grandparents, and we knew each other well. I was also old enough to know my maternal great-grandparents.

My maternal grandfather, Rev. Joseph Henry Cashen, worked a full-time job at a lithograph plant in Baltimore. Granddaddy, as we called him, didn't drive, and I often drove him to his overnight shift work. My maternal grandmother, Marie

L. Cashen was a Red Cross volunteer who did all the driving until her legs were amputated due to diabetes. I remember her volunteering and participating in service projects and events a couple of times a week for years.

My paternal grandparents, Richard Ward, Sr. and Mary Parker Ward Flood, were divorced. He was a handyman, and I would spend time with him and his friends as they did odd jobs in and around Baltimore. However, as with many Black families who'd migrated north, I'd spend summers with my grandmother in Bedford, Virginia, my father's hometown.

During World War II, Bedford, Virginia lost more men per capita than any other locale in the United States. Staying in the house where my dad grew up, I learned a lot. It was a humble home, with a well outside for drawing water daily and an outside toilet. Here was where I learned what it meant to be in a place without running water and using an outhouse for the bathroom. It was humbling and reinforced my awareness to not take things for granted.

I was blessed to also have known my maternal great grandparents: Claire Cashen (Grandma) and Richard Cashen (Grand Pop), who had served in the Army in World War I. Grand Pop was a part of the legendary New York outfit, the Harlem Hellfighters, the 369th Infantry Regiment. This was an all-Black unit, serving under mostly White officers. By the end of the war, the Harlem Hellfighters, (most likely given that name by their German adversaries), had spent more time in continuous combat than any other American unit of its size, with 191 days in the front-line trenches, according to the National Museum of African American History and Culture.

These soldiers were willing to join the Army, thinking their service would lead to better treatment at home. However, segregation was alive and well in and out of the military and remained so once they returned home. Grand Pop never talked about the experience, but I know from Granddaddy and Ma Marie that he was gassed in Europe, exposed in trench warfare.

He was treated for "shell shock" in a Pennsylvania veterans' hospital upon returning from overseas. Today, what he suffered from is called post-traumatic stress disorder, PTSD.

Those summers in Virginia were an absolute joy—exhilarating memories of riding bicycles along country roads and being with friends, including time with my cousins on my dad's side. The oldest girl and her husband had a farm and raised horses. I would spend time with them in Lynchburg, Virginia. Summer days were filled with working on the farm, bailing and stacking hay and doing chores of that nature. Although I am sure that it was around me, I didn't feel racism. The communities were separate, and because of my age and interests, it just wasn't something that I was aware of.

During the time I spent with Grandma Mary, I learned what going without something meant. This experience added to my appreciation of things that I had. My grandmother didn't complain about it, but I knew managing day to day was something difficult for her. But somehow it all worked, from cooking on a wood-burning kitchen stove to taking a bath inside in a large tin tub.

My childhood experiences in Baltimore city and in rural Virginia with family and friends were just wonderful. We weren't rich by any means. We didn't starve, but at the same time I knew there was another world going on around me. I never felt bad or disadvantaged because we had what we needed. I was able to work and do things that helped me see down the road a bit and have a vision.

School Days

My early education began in Baltimore. My first school had no kindergarten or preschool. My preschool days were spent with my maternal grandmother, Marie L. Cashen—we called her Ma Marie. Being with her as my parents worked was a true education in being honest, respectful and courteous and understanding the consequences when you weren't.

In Baltimore, 1955, I began first grade in a segregated, two-room school building school in our Black community. It had 1st – 3rd grades in one room, and 4th – 6th grades in another, and two teachers. The 4th – 6th grade teacher was also the school principal. After the 1954 Brown vs. Board of Education Supreme Court decision, schools made plans to integrate in Baltimore, so I only spent my first-grade year in a segregated school.

I went to Medfield Heights Elementary School, an integrated school, for my second-grade year. The school was near our little community, so I walked about a mile to the brick and fully outfitted school. An impressive school that had been in place always for white kids now was open to Black students. I attended that school for four years, 2nd through 5th grades. By the time I entered the 6th grade, my parents had finished building our home and I went to Rodgers Forge Elementary, another integrated school in Baltimore County. I was the only Black student in my 6th-grade class.

Being the only Black student was unsettling for me, especially during one particular grammar lesson. Verbs were the topic, and I incorrectly answered a question about helping verbs in relation to adverbs. The teacher blurted out that I needed a helping verb. I left that class still not fully understanding helping verbs. Others made mistakes as well, but none evoked the reaction I received from the teacher. Later, telling my mom about the incident and never wanting to be embarrassed like that again caused me to buckle down and make sure I understood my lessons. I was being treated differently and had to be better. That lesson I never forgot.

Growing up in an all-Black Baltimore neighborhood of blue-collar workers, attending a small church pastored by my grandfather and living with my grandparents as my parents built our home made a great impact on my psyche. As I watched my dad build our house because he could find nothing worthy for his family, I interpreted him saying: I'll take care of this problem myself. This informed and confirmed a lot of the decisions I would make in the future. My mother's supportive actions and discerning vision—while working outside the home every day to help make ends meet—fed my faith and confidence.

My parents pooled their resources, worked together and raised my sister and me. My grandparents were big time supporters along the way. I spent a lot of time with my grandmother and observed her volunteer spirit. I watched my granddad work and pastor a church.

I spent time in Anne Arundel County, Maryland, where my paternal grandfather lived. In the summers I was in Bedford, Virginia with my paternal grandmother. I saw their interactions with other people and how they reacted. I watched my father treat everyone with respect and dignity. I often think how far he could have gone if he had the opportunities that I have had.

Not long after I graduated from college and on a visit back home after entering the Army, I found the architectural designs that my father had drawn for the house. He didn't go to an architectural firm, but drew his own design, and then followed his own blueprint for building the house.

I watched my mother work every day. In the early years after getting married, she worked in the office of a Baltimore chemical company. Eventually, she would move on to work as a filing clerk for the Social Security Administration retiring as a GS5. Mom and Dad provided for my sister, Christina, and me and gave us the things that they didn't have. They were committed to providing for us. Over the years as we moved forward, my sister and I were able to repay some of that.

During this time, my parents also demonstrated that life was also about living. Ike and Phyllis Ward were active in multiple houses of their masonic lodge and Eastern Star organizations. My dad was a 33rd degree Master Mason and Shriner. Both held leadership positions in their organizations. They found time to have fun with sports and social organizations. I'm convinced my dad could have been a major league baseball player; he was that good. I was a catcher in high school because my dad was a great pitcher. I was his warm-up before his Westside Eagles ball games, so you can imagine that I got pretty good.

I attended Towson Senior High School in Baltimore County and was an above average student athlete, participating in varsity sports: football, track, and baseball. In 1967, after graduation, my mom and dad said that I was going to go to college in Baltimore. At that time, only one member of my family on either side had graduated from college and that was my mom's baby sister, Doris. (Yes, the one who named me Kip.) She had graduated from Morgan State College, so my family knew about Morgan.

Interestingly, my counselors at the integrated yet predominantly white Towson High School offered no additional insights into other options for college nor did they suggest applying for scholarships. My whole college application process occurred without help from the officials at Towson. Today, high school teachers, counselors and leaders are encouraged to help students seek every opportunity and to instill hope in each child.

Although I wasn't the class valedictorian, I was an above average student and varsity athlete, lettering in football and track. Had things been different (equitable), I would have, at least, been presented with other options to pursue.

Going to Morgan State was an identifiable step to be taken for my generation on either side of my family. For us, Morgan was the perfect place to attend, ironically at the height of the Vietnam War. For me and my family, certainly Morgan mattered. In Baltimore, it was the school that had a reputation of providing for and caring for Black students.

Morgan is what made sense to me, and at Morgan I was accepted. Determined, I entered with a goal of graduating.

Morgan and ROTC, Here I Come!

Like many, when I graduated from high school in 1967, I was looking at my options for the future. The war in Vietnam was going on. In many ways, going into military service seemed almost natural—at least it rang familiar in my family. I thought about my dad who was a World War II veteran, a combat engineer who had served in the Pacific theater. I still recall being with him on visits with a few of his Army buddies who returned to live in Baltimore. He didn't talk a lot about it, but I heard stories of his service and knew he would be happy to see me in uniform as an officer. I certainly knew he served with pride and respect. I thought about my uncles who served in the Korean War.

In 1967, Vietnam was in full force and people were being drafted—two of my cousins on my mom's side were drafted, and a third on my dad's side had enlisted. William Burwell, Jr., wound up being a careerist (a career soldier); he retired from the military. The other two cousins, Lawrence Gross and Thaddeus Price, returned from Vietnam, served their draft enlistment, and then left the military. They are no longer with us, but like my dad, both were buried with military honors. I was aware of other service members in my family and none of them ever said anything negative about the military. I reasoned that if I was going to Vietnam, I'd go as an officer and staying in Reserve Officers Training Corps (ROTC) was the ticket that afforded me that opportunity. I would graduate with my diploma and a commission.

Additional awakenings began to happen now. We had lived through the killing of Martin Luther King Jr. and the riots in 1968 and the aftermath and reverberations of the Civil Rights Movement. During that time there were Vietnam protests, but the ROTC detachment at Morgan and what we did on campus was never subjected to harassment. We were still doing what we needed to do and were respected on campus. We participated in various activities, from flag raising on the campus to performing our drill and ceremony exhibitions.

Civil protesting was a different experience for us, unlike what was happening at some other schools. In the spring of 1968, we were all coming back from a cadet field training exercise (FTX) and coming through Baltimore, hearing and seeing the rioting.

Martin Luther King's death and its aftermath happened during that time, but the fallout from it did not impact those of us who were in ROTC or our decisions. I'm sure there were some who may have decided to take other routes, but many of us stayed. That was significant for us due to the support that we continued to receive from the school and community.

At Morgan State, like many other land grant and state colleges in those days, two years of Reserve Officers Training Corps was mandatory for male students. I enrolled in the ROTC program and spent my freshman and sophomore years taking ROTC training, embarking upon a degree in political science and other courses necessary for graduation.

My freshman year went successfully, and I was elevated to a sophomore and matriculated through my ROTC as well as my regular academic studies.

After ROTC's mandatory 2-year requirement for male students, a person could leave the program but then face the possibility of being drafted. After my two years, I decided to stay in ROTC and graduate. If I were assigned to go to Vietnam, I'd go as an officer. So, for my remaining college years, I majored in political science with the goal of going to law school after my military commitment.

My professors there, Drs. Gill, Walker and Adair, were political science faculty who were esteemed gentlemen in my mind and impressed me. They were distinguished men. My professor of military science, Colonel Clarence Mel Hurt, was the ultimate military professional, a true example of a soldier to emulate, and we cadets did. Captain Nicholas Lagatutta and Sergeant First Class Broadus Mattison, two other cadre members, were also very instrumental to my development.

As an HBCU (historically Black college/university), Morgan had a reputation for holistically taking care of its students. Morgan lived up to that reputation—for me it was not only enriching and educational, but enjoyable. I learned while

I worked through school. I also did things that were important to maintaining my family relationships with my parents and extended family. I made time for family gatherings. To save money and because I lived in Baltimore, I didn't live on campus; still I had an active campus life and got to spend a lot of time with many of my classmates.

I worked all through college. Even though my parents provided for me, I knew that attending Morgan came at a major cost for them. Not having saved a college nest egg, my folks paid as I went, and I wanted (needed) to help. Although in those days the tuition, fees and books weren't what they are today, for us it was still something to be planned for and worked out. I knew that my parents were the main drivers. It wasn't as if we had saved up for college and my sister would enter Morgan my junior year.

We were paying for my education as we went along —sort of like they did with our home. Seeing an opportunity to help, I applied for and received a two-year Army ROTC scholarship. My parents understood the ROTC scholarship I had received came with a four-year, active-duty commitment and my mother especially was not enamored that I accepted it. But I knew it would relieve a financial burden. My sister was also about to enroll in Morgan, and they would have two children enrolled in college.

I participated in campus student union and other social activities. My role in ROTC gave me a lot of involvement in campus life; that program began to define some things for me. I knew that I enjoyed being amongst my fellow cadets and looked forward to participating in my ROTC program.

The cadre in the ROTC department saw something in me and put me in charge of various exercises. Those first couple of years I enjoyed being a part of that team and then coming to lead that team. For my first couple of years in the cadet corps, I was not only a member of the squad, but also a member of a fraternity called the Pershing Rifles. We had a drill team and traveled around performing basic and trick drill marching routines. I got pretty good at handling weapons and proficient as a member, so later I was named commander of the drill team.

We performed at halftime for the basketball team in Baltimore at the Civic Center and at many other community gatherings and celebrations. I was drawn to and enjoyed the training exercises, the leadership labs, as well as the academic courses emphasizing effective leadership. All the while, the activities that fit me so well were defining and shaping my characteristics and behaviors.

Although ROTC was no longer mandatory for my third and fourth college years, I elected to take the scholarship and stay in ROTC. By then I understood what it was about (to include a draft exemption allowing me to graduate) and frankly it was something that connected me closer to my personal purpose.

Meeting Joyce

When I started at Morgan, I had no idea that the people I would meet would not only impact my life but also would be in my life in a major way for decades to come. That reality began when I first met Joyce Lewis, another student just starting at Morgan. I saw her often because we had freshman classes together, French and social studies. We recognized and knew each other's names.

Joyce chose Morgan after winning a partial scholarship. She was from Virginia, had lost her dad at the age of eight, and had an older sister who lived in DC. We started dating at the end of our sophomore year. After a conversation on the campus bridge, we became a bit more serious. We continued dating that summer and throughout our junior year.

Things began to take shape for me with respect to my grades and performance at Morgan. That scholarship required good academic standing and thankfully I was leaning into that as well; I was both a cadet leader and successful political science major.

Joyce, who was smarter than me (and still is), was one of Morgan's few women majoring in business who completed their basic program in three and half years. For her final semester, she would do an intern program in the DC government.

ROTC is loving me, and the feeling is mutual. As a junior, I assumed a leadership position in the program. We would go on weekend training field exercises to places like Fort Meade, Maryland. I led various tactical activities and drills and liked being in charge. I enjoyed leading those activities and frankly I enjoyed winning, being a part of a team that was winning, and leading in a way that made a difference.

For my third and fourth year, I earned several ROTC commendations. At the end of junior year, our cadet summer camp training was held at Fort Indiantown Gap Military Reservation (IGMR), Pennsylvania. We went there and linked up with ROTC cadets from all over the mid-Atlantic, Northeast and parts of the southeast and into the Midwest from other universities. My ROTC cadre from Morgan along with the professor of military science prepared us to compete in that environment.

We went to that summer camp at Fort Indiantown Gap in increments or cycles. The summer camp was a six-week program. We did well there, and my performance as a top cadet netted me the position of serving as the brigade commander for my increment at graduation.

I realized and understood fully that I was being prepared well for some of the basic skills required to be a soldier and an officer. Noncommissioned officers (NCOs) worked with us, and we knew how to conduct ourselves as a part of a team. Duties would rotate. The training included assignments to various leadership positions. On any given day you might be a member of a squad or platoon; and on another day, you would take leadership in that squad or platoon as squad leader, platoon leader, platoon sergeant or fire team leader. All while this was going on, we were being evaluated by the summer camp cadre, typically ROTC instructors from the various schools. They were part of the summer camp cadre training the cadets.

We navigated through summer camp, and with graduation also came the decision to select the graduation brigade commander. Many factors were considered for which cadet would be the brigade commander. The person selected would be someone who had served throughout the camp: doing all assigned duties, doing them well, and excelling when given leadership positions. Conducting all tasks well and having demonstrated sound leadership through a strong command presence were the attributes sought in identifying the commander. That would be the person to serve as brigade commander for the graduation.

For the graduation ceremony, this brigade commander would oversee the formation. Other cadets were selected to fill leadership positions, including battalion commanders, company commanders, platoon leaders, and squad leaders. Cadets would be selected to fill these various leadership positions across this entire cycle of camp. Then the selected cadet would lead a brigade-sized formation; a brigade was typically 3 battalions with a battalion containing typically 3 or 4 companies with 3 or 4 platoons in a company, with squads in each platoon in the number of 3 or 4. With a pyramid type of an organizational construct, the brigade commander is at the top.

From the various activities conducted, I—Cadet Colonel Kip Ward—was selected to be the brigade commander for my summer camp cycle's graduation! That was a substantial achievement. For increment #2 at IGMR ROTC summer camp graduation, I received the nod.

Being selected as brigade commander was a big deal. I didn't know how much it meant until my professor of military science, Colonel Hurt came and told me that it was a notable achievement, "This is a big deal, and you've made Morgan proud by being your cycle's brigade commander."

Unexpectedly, that opened a door because at the conclusion of summer camp, when I went back to Morgan for that senior year, I was also selected to be the brigade commander of Morgan's ROTC brigade detachment. I was the senior cadet at Morgan and the captain of my Pershing Rifles Company. Another significant achievement! The Pershing Rifles fraternal organization exists on many college campuses but originated in the ROTC departments with respect to ROTC cadets that were matriculating in colleges and universities. Now, I was also the commander of Company H-15 (now H-8) Pershing Rifles at Morgan State College (now Morgan State University).

As my ROTC professor of military science, Colonel Hurt was just an impressive, impressive officer! The senior drill instructor was Sergeant First Class Broadus Mattison, a senior NCO and an absolute professional.

Colonel Hurt was a man for whom I have deep admiration, specifically for his professionalism, his care, who he was as a man, and as a soldier. He is still the standard by which Joyce and I judge a professional soldier. Between him and my dad, who was not a career soldier but a professional who always displayed teamwork and respect, both had a profound impact on me.

An assistant professor in the ROTC department, Infantry Captain Nick Lagattuta, spent a lot of time with me, giving attention and support. These men were professionals, and they saw something in me that I didn't see in myself at the time. They worked with me, encouraged and pushed me to achieve things that sometimes I didn't know I could achieve. But they saw something and pushed me to do those things.

In addition to my ROTC instructors, my political science professors were likewise great role models. Drs. Gill, Walker and Adair were influential and led me to consider a career in law.

That June, just prior to graduation commencement, our class was commissioned. The commissioning officer was then Brigadier General (one-star) Chappie James, an Air Force officer who later became the nation's first Black four-star general. He was a famed Tuskegee Airman and fighter pilot. I didn't realize how significant that was for me to have this officer commission my class.

In June 1971, 15 of us were commissioned from Morgan into the Army. As the commissioning speaker, General James oversaw the pinning of one 2nd Lieutenant bar on one shoulder and Joyce (by then we had been married about two months) and the other by my mother. My dad and sister and Joyce's mom and her sister and brother were there. It was a heady, heady experience that I will always cherish. In addition to graduating with a bachelor's degree in political science, I was commissioned and graduated as a distinguished military graduate (DMG). That commission was a regular Army commission, meaning upon graduation, I would be going on active duty as a regular Army officer. And by the influence of Colonel Clarence Hurt, I had been branched as an infantry officer.

From University to Army Life

For the class of 1971 at Morgan State, graduation ceremonies happened during the first week in June. Although I was still attending class during my final semester, soon I would report to the Army for my first duty assignment. So, June 6, 1971 (D-Day) was a significant day in our family history, graduation day, and going immediately on active duty with a 6 June date of rank. I was to report to my initial duty assignment on the 8th of June—exactly two days after graduation.

By the time all of this was unfolding, Joyce and I had declared our love for one another and wanted to spend the rest of our lives together. After consulting with my parents and her mom, we made the decision to get married, and because I was going to leave on active duty, we would get married just prior to graduating.

We married on the final break day of our senior year, April 3rd, in the spring of 1971. Our honeymoon was an abbreviated trip to the nation's capital because we couldn't afford anything else, and we had graduate record examinations the following Monday. When we made our wedding plans, the exams had not been scheduled.

Joyce had completed her coursework in 3.5 years and got that intern position with the DC government. She was earning an income! Her income afforded us an opportunity to get married so that upon graduation (Yep! Notwithstanding my part-time job and $50/month ROTC allotment) I had been a kept man for two months. We entered military life as a married couple, sticking with the plan that in four years I'd get out of the military and go to law school. Later Joyce joked, "He only asked me for four years, but then it wound up being four decades."

Two days later, I was on my way to Fort Bragg, North Carolina to join the 82nd Airborne Division as a brand-new 2nd Lieutenant of Infantry. We didn't realize it so much at the time, but that was a significant event for me and for our family.

CHAPTER 2

Military Blood Flows Vein Deep!

It's 1971—I am a confident 22-year-old with the strength of an HBCU education on my shoulders. I'm a newlywed with military blood already flowing in my veins, eagerly anticipating Army life! My new future is all I can think about on my way to my first assignment, Fort Bragg, Fayetteville, North Carolina.

Another lieutenant who had graduated from Morgan and I were driving in tandem on U.S. Route 301 (I95 was not complete.) He was driving behind me. As we exited off 301 heading towards Fort Bragg, I saw this big billboard displaying a Ku Klux Klansman mounted on a horse, with the words: *Welcome to the Heart of Klan Country, join the United Klans of America.* I was absolutely flabbergasted.

The billboard was outside of the post, leading into the city of Fayetteville. Gradually and cautiously, we just slowed down and pointed up at the billboard as we passed it, wondering *"Where are we now?!"* We both realized it was our first encounter with this physical public display of Klan mentality. We could see first-hand what was going on with the United Klans of America. Stunned, we drove on to report to the 82nd Airborne Division.

We began our journey as new lieutenants. I was assigned as a platoon leader in Alpha Company 3rd Battalion, 325th Infantry. It was the beginning of a journey that I had no idea would take me and my family through the next 40 years. That journey began on Ardennes Street, Fort Bragg, North Carolina, in the 2nd Brigade (the Falcon Brigade). I just melded right into that organization.

That very first stop at Fort Bragg was only going to be for a couple months because I had not been to basic branch school.

That would be about 8 hours away, at Fort Benning in Georgia. The post is known to provide "trained and combat-ready soldiers and leaders, to develop doctrine and capabilities for the maneuver force."

First flight, first jump!

Being assigned to an airborne unit meant I had to attend parachute training, in addition to my infantry officer basic course. I'd never flown in an airplane prior to airborne training. After two weeks of ground and jump tower training, I had my first flight, and I jumped safely. I did not depart the plane like most passengers flying on an airplane for the first time; I jumped out after taking a perfect door position, a slap on the back by the Jumpmaster, maintaining a tight body position while waiting for the opening of the parachute and checking my canopy overhead, then drifting down to the ground to execute a safe five point parachute landing fall (PLF!).

Becoming a paratrooper through the Army's jump school included completing five parachute jumps to graduate. And uniquely, the planes were from the Air National Guard. They were close to ancient Fairchild C-119's, affectionately known as Flying Box Cars. They were so old and loud, we used to joke that we could see the rivets bouncing on the fuselage. So, we were glad to jump out.

That experience of jumping from an old, loud airplane in flight into the quiet of floating through the air with a parachute canopy above your head—once you opened your eyes—was an experience of a lifetime. It was short lived, however, as the jump occurred at about 1200 feet above ground and the landing—either a hard hit or a perfect execution hitting all five points of contact—came fast.

What I thought could be my final jump occurred when I left the 82nd Airborne Division as the Assistant Division Commander in 1998, some 27 years later. Part of the farewell included a jump from a Douglas DC-3, which was a plane used in World War II by the paratroopers jumping on D-Day. That was a special jump as I approached that platform. It wasn't the hardest landing I'd ever made, but it was a windy day, and I was late preparing to land as I savored the descent from the DC-3. Somehow, the hard landings, especially with equipment bags and weapons tied to your body, are intimately remembered.

Once I'd completed the basic course and airborne training, I returned to my unit in the 82nd Airborne Division. At Fort Bragg, I led a platoon of soldiers, many of whom were veterans of Vietnam; they were non-commissioned officers and soldiers who had been drafted in the past 5-7 years, and several were ready to get out of the Army. That experience was especially formative for me—being in the 82nd Airborne Division, known as America's Guard of Honor, always ready to deploy.

I took that assignment and my responsibilities seriously. I wanted to be a good platoon leader. Morgan's ROTC program had prepared me well. I paid attention to my sergeants. I knew that was crucial for being a good leader. How I interacted with the non-commissioned officers mattered. I listened and took advantage of their experiences to help me lead. While I was the lieutenant, the leader in charge, I understood that being in charge did not make me better than anyone else.

Merging Marriage and Military Life

Joyce had started an internship with the government prior to our graduation from Morgan; however, she joined me at Fort Benning as I began my studies at the infantry school. We moved back to Fort Bragg together. This was a wonderful experience for us—making major life decisions together. We moved into our first house, military housing—a two-bedroom home, part of a row house complex on Fort Bragg called Corregidor Courts. This was now our home. Although after our April marriage, we lived for two months in a small apartment in Baltimore, the place at Fort Bragg would be our first home where we would begin raising our family.

Years later we mused over how small it was, but at that time, it was our mansion. We enjoyed furnishing it and living there. We acclimated easily into the stride of married life. I was working every day going to my unit area and training. Joyce looked for employment and ultimately chose substitute teaching, mostly at Fort Bragg schools. (The military has since come a long way in supporting professional careers for spouses.) We enjoyed our friends and our social life. Many of the military couples we met then remain close friends today. Our military families are not family by blood but are just as close—military close. These relationships are not planned, they happen because of the environment. Despite that KKK billboard "welcome" sign, ours was a productive and valuable tour at Fort Bragg, the home of the Airborne.

Early in 1972 we were blessed with the news that Joyce was pregnant. The reality of a first child brings a watershed of feelings, flooding one with responsibility for a new life. Although young, we were overjoyed at the news of having a child. I began imagining our life with a child and knew we'd need another car; my beloved Mustang had to be traded for a family-friendly car.

When Joyce went into labor and we went to the hospital, we thought her delivery time was just beginning—assuming we'd be there for a while. I'd planned to be with Joyce the entire time. As we were waiting between contractions, the doctor said it would be a while and suggested I go downstairs to the vending machine and get a snack. I was only gone a few minutes, and when I returned, Joyce had been wheeled into the delivery room. In the few minutes that the nurse had come to check her, Joyce was ready to deliver. The next thing I knew we had a son! That doctor was sure her labor would be longer, but he was wrong. Looking back, I still wish I'd gotten the nurse's opinion before I left to go to the vending area.

We celebrated our first-born son, and named him William Edward Ward, Jr. We'd been thinking about it a while and mused that, boy or girl, we'd call our baby KJ—taking the K in Kip and the J in Joyce. Because KJ is male, some friends equated the initials to mean Kip Jr., but that is not the case.

Just after KJ was born, our little family made a weekend trip to Baltimore from Fort Bragg. Once we were all together, we realized there were five generations of men together: Richard Cashen (maternal great grandfather), Rev. Joseph Henry Cashen (maternal grandfather), Richard Isiah Ward, Jr. (father), William E. Ward (me), and KJ, William E. Ward, Jr. (my son).

Gathering in the parsonage after attending Granddaddy Cashen's Sunday church service, the five men posed together for a photo. That was really a special deal. My son had been born in the fall of 1972 and here I had a great-grandfather who was still living. Over the years that photo disappeared, but that experience and its memory are strong; it's something I'm very proud of and thankful for having.

Fort Bragg—the Early 70s

My experiences as a lieutenant continued to expand—beginning as a platoon leader in A-Company, 3rd Battalion, 325th Infantry (Blue Falcons); then becoming an executive officer, followed by being a specialty platoon leader in Headquarters Company, 3-325 Infantry.

There were three specialty platoons in the battalion: mortars, scouts, and anti-tank. Mine was the anti-tank platoon with 106mm guns mounted on M151 Jeeps. I led that platoon long enough to field a new weapons system. That was an anti-tank guided missile system that the 82nd Airborne received named TOW (tube-launched, optically tracked, wire-guided missile). My platoon became one of the first units to receive that specific weapons system, mounted on a "mule," a flatbed all-terrain vehicle with no protection, making crews vulnerable to return fire. Employment tactics had to be very precise. The rounds were extremely expensive. Live fires were limited and most training was via simulator. My part in that fielding process gave me valuable background for subsequent new equipment fielding programs I would experience.

At Fort Bragg, I refined my understanding of what it meant to be a platoon leader. Through the exercises and training drills, I picked up a mental technique, almost a special sense, that prepared me for weathering and recognizing upcoming issues/ challenges, developing lessons that became trademark practices in my career. The lessons I learned from Sergeant Mattison and CPT Lagatutta at Morgan were clear, well-received and essential to my foundation. I will always be grateful for that.

Despite that negative welcome sign, my time at Fort Bragg—my lieutenant years—provided huge growth and learning opportunities. I had every job a young infantry lieutenant needs to experience:

- Rifle platoon leader,
- Company executive officer,
- Specialty platoon as anti-tank platoon leader,
- Battalion staff as battalion motor officer (BMO), and
- Brigade staff as brigade assistant operations officer (S3).

One day something unexpected happened. I was tagged to get the job as the battalion motor officer (BMO). The assignment brought back memories, stories I'd heard about the types of jobs that Black officers get, jobs that do not lead to higher command level positions. This was one of those jobs. Again, this pattern would raise its condescending head, *"It's your turn and you're going to get one of these jobs."* But the question was: why did I, as the only Black lieutenant in the battalion, get this job. The ROTC guidance model I was taught as a Morgan cadet resonated: *Do the job that you are assigned, that you are told to do. Do that job to the best of your ability and other things will take care of themselves.*

Many assignments given to Black officers were at battalion level, such as a motor officer; those were the jobs that Blacks got. I knew that, but it was the job that I was given, and I was going to make the best of the situation. So, I asked to go to Fort Knox (Kentucky) for the maintenance officer's course. If I was going to be a motor officer, I needed to understand what I was going to be responsible for and gain a better understanding of the system. The unit supported my request for this training.

During this time there were two job openings for lieutenants in the battalion: the battalion motor officer and the support and transportation (S&T) platoon leader. Platoon leader was perceived to be a more significant position than that of motor officer as that lieutenant was also considered the assistant logistics officer in the battalion (Asst. S4). Not surprising, a white officer was assigned to the S&T platoon. That told me I was going to be the motor officer.

I rolled many questions over in my mind: *Why didn't I get the S&T platoon job? Why wasn't the other guy assigned as the BMO?*

My assessment of my chain of command's answer: *That's where guys like you go.*

My reaction to it all: *I'm going to show them. I will go to this maintenance course at Fort Knox and I'm going to learn the ins and outs of maintenance, procedures, systems management, and readiness; I was determined to master every skill and technique required. My goal was to excel like nobody ever had excelled at this position.*

So, in 1973 Joyce, KJ (barely a year old), and I packed up and traveled together to Fort Knox, Kentucky. It was a 90-day course, but I brought my family with me. We were authorized to take *our* families on this temporary duty (TDY) assignment. That set a dynamic for us that was sustained throughout my career: *Whenever possible, we would be together as a family.*

I took full advantage of this maintenance course. Studying, training and learning about engines, systems, and maintenance readiness prepared me to come back to the battalion as the motor officer. My training went well from a maintenance perspective and I developed an understanding of how to keep equipment maintained and in a high state of readiness. Had I not taken that motor officer's course—a course normally reserved for armor officers, my performance would have suffered because at that time I lacked that knowledge; not taking that course could have led to my detriment in the future.

When we returned to Fort Bragg, I made history in upgrading the maintenance status of the Blue Falcon vehicles. My newly learned maintenance expertise coupled with being the type of leader that my soldiers respected because they knew I trusted, respected, valued and took care of them produced the highest level of maintenance readiness the battalion had seen. The 2½-ton trucks (called deuce and a halves) —used by the support platoon to carry paratroopers and supplies—had a high operational "ready rate" and were at their best in taking care of soldiers. What a fantastic morale boost (and readiness enhancement) when the soldiers knew they could count on transportation, food and other supplies being on time and on point.

I was not excited about this assignment often given to Black officers, nevertheless, not only did I do the BMO job, I excelled at it. I didn't know or understand it then, but that experience paved the way for my future success. With almost four years on active duty and in the window for promotion to captain, the next presumed career step for senior lieutenants was to return to Fort Benning for the infantry advance course.

However, just prior to reaching that time and before my four-year active-duty obligation hit, I was on permanent change of station orders to leave the 82nd Airborne Division and report to a heavy (mechanized infantry) battalion in the 2nd Infantry Division in Korea.

The notion of using these foxhole analogies kept surfacing. I began to mentally collect these lessons—unique aspects of my journey and elements that defined my life—that ended up benefiting me later. Those foundations began with my great grandfather, WWI Veteran Richard Cashen. They directly related to experiences I gained in becoming a leader, some were behaviors, some were simply understandings and confirmations. Along my career, I kept noting foxhole lessons.

> ***Foxhole Lesson: Accept assignments—you never know how that experience may inform and transform you for future challenges.***

Due to the real-world missions of the 82nd Airborne Division, there were times when I was often separated from my family. In 1973, my unit was at Pope Air Force Base (North Carolina), waiting for directions. We called the staging area for short notice quick deployment missions the "*green ramp,*" it was the paratrooper holding area at the departure airfield where the units prepared to conduct an airborne operation and waited until time to load the planes. That wait could be minutes or days, depending on the mission. In the case of the Yom Kippur War, my battalion, the 3rd Battalion, 325th Infantry (Airborne), was on alert and we were standing by in case President Richard Nixon ordered us to go to Israel to support their defense.

As it happened, after a few days on alert, we stood down and never deployed to the Middle East. That was the first time it really hit me that being an Army paratrooper in the 82nd Airborne Division was a big deal. We had to be ready to go anywhere on a moment's notice. Our training in the 82nd, both local training and exercises as well as overseas deployments to places like Panama for Jungle Warfare, kept us ready.

As I was nearing the end of my tour in the 82nd Airborne, the commanding general of the Army's 2nd Infantry Division in Korea sent an urgent message to the Department of the Army saying they needed senior lieutenants in the division. At that point the general was getting newly commissioned lieutenants and captains coming out of Vietnam. Missing was mid-level junior leadership, experienced lieutenants who had been platoon leaders and could provide effective company level leadership in the division.

Also in 1974, the Army (as was the U.S. society) was experiencing racial tension within the ranks. There were racial encounters between Black, White, and Hispanic soldiers that included blocking entrances to dining facilities and clubs; clubs divided according to race. Dapping was often used as a public display of identity. It was seen as a sign of unity for Black soldiers, mirroring what was happening in the general population (albeit to a lesser degree) during the struggle for Civil Rights.

Infantry Officers Career Course

In lieu of attending the career course, the Army assigned a number of senior first lieutenants to the 2nd Infantry Division in Korea. We were sent from divisions like the 82nd Airborne Division, the 101st Airborne Division and the Third Infantry Division; a number (30-40) of first lieutenants found themselves in Korea instead of returning to attend our career courses.

My Thinking: Leave Fort Bragg (North Carolina), go to Fort Benning (Georgia), for the advanced course—a six-month school to be followed by company command. I had a four-year, active-duty obligation considering my regular Army commission and the 2-year ROTC scholarship. My four years hadn't quite hit, I had about 3.5 years of service. I was still obligated to serve active-duty time.

Army's Thinking: Ward goes to Korea on a one-year unaccompanied tour for assignment to the 2nd Infantry Division's 1st Battalion, 17th Infantry, the Buffaloes, a mechanized infantry battalion.

I would not be part of an airborne unit, jumping out of airplanes as in the 82nd Airborne Division, but headed to a heavy (mechanized) outfit in the 2nd Infantry Division. As I processed this change, good news appeared: Because of that motor officer's course at Fort Knox and the BMO job, I knew about maintenance, about the equipment, and the systems and readiness issues. My previous assignment had prepared me for a job I had no idea I'd be doing. I didn't know it at the time, but this was a foretelling of things to come.

Sometimes it's difficult to understand the reasons for a specific assignment. You wind up in seemingly strange places. Then, down the road you'll be asked to do something, and that previous job prepared you for what's ahead.

> ***Foxhole Lesson: Critical—You never know what is being set up for you, so take advantage of every opportunity.***

Arrival in Korea

A tour in Korea was not something I expected! I had no idea it was coming, but it came my way. Because I was still under my active-duty obligation as a regular Army (RA) officer, I didn't have the luxury of declining that assignment.

As it was an unaccompanied tour, Joyce and KJ couldn't come with me. We made the decision that they would go to Baltimore and spend time with my family, giving Joyce opportunities to expand her career. So, Joyce left Fort Bragg with 2-year-old KJ, primed for enjoying and being spoiled by his grandparents in Baltimore. She could be around family there, and so I embarked on my solo tour to Korea.

Upon arriving in Korea and reporting to the 2nd Division, I was further assigned to a mechanized infantry battalion. How convenient since earlier, I had been sent to the battalion motor officer school and completed the training required for this mission. It was a mechanized infantry rifle battalion operating along the Korean Demilitarized Zone (DMZ). I was assigned to B Company of the 1st Battalion, 17th Infantry. Coming from the 82nd Airborne, my thoughts were: *A few months ago, I was a paratrooper jumping out of airplanes. Now I'm going to a mechanized infantry company with these mechanized vehicles, M113 Armored Personnel Carriers.*

Instead of this being a recipe for failure for an airborne infantryman, that battalion motor officer school prepared me for understanding the systems, materials, the diesel engines, and the heavy equipment. Although we didn't have all that equipment in the 82nd, the school had that equipment as a part of the program of instruction. In addition to my tactics as an infantryman, I had maintenance and management understanding of these systems going for me as well.

I had no idea when I was going through that program that it would now come in so handy and be so valuable. My initial duty in Korea in the 1st Battalion, 17th Infantry, 2nd Infantry Division was serving for about two months as a special assistant to the battalion commander. I performed troubleshooting and management of various activities and processes.

After spending a brief time as a company executive officer in B Company, I was selected to Command C Company in the battalion. I was still a first lieutenant and had not been promoted to captain.

A Jarring Reality

During this Korean tour, early in 1975, while I commanded Charlie Company, one of my squad leaders had a coronary incident and died. It happened during physical training (PT) that morning. Although it was not a combat loss, for me the loss landed heavily. Our PT—running, strength training, stamina building, among other things—was intense. He was a staff sergeant, a good soldier with a family and now, in Korea, away from his family, his life had ended. The death of a soldier is something you know can happen but cannot prepare for.

Dealing with grief from the loss of one of my soldiers was like losing a close family member. Although the Division's Casualty Assistance Office provided support and was primarily responsible for handling matters, I wrote letters, planned arrangements, contacted casualty assistance officers, and assisted the family as best as I could—in that far away environment. The letters and expressions of sympathy told the family about the type of person he was. I wanted them to know that he served honorably and well.

This was my first experience dealing with the ramifications and impacts of losing a soldier. The soldiers in my company and the deceased sergeant's soldiers were right there in the unit. So, providing bereavement support for his family and the people under his command became my focus.

That harsh reality in Korea handed me the experience of determining how to reach out to a deceased soldier's family and how to do all within my power and resources to ease that family's burden. It would not be the only time I would use that new knowledge.

As I said, our cohort of first lieutenants was a sought-after group. My battalion commander took special note of me. The 2nd Division, like the rest of the Army, was experiencing racial issues. Supply accountability was almost a joke because there was very little emphasis on accountability: often equipment was stolen or lost and not accounted for properly.

Around the three-month mark of this one-year tour—and after tackling some important readiness, unit cohesion, and training problems—my battalion commander informed me that he and the brigade commander decided to put me in command of a mechanized infantry rifle company. I was still a first lieutenant and had not been promoted to captain. Normally captains command companies, so this would be a significant step for me. Getting to command an infantry company in an active theater of operations—the Korean DMZ—was a big deal. The kick was that if I commanded the company, I would have to extend my tour and I was already three months into my one-year tour.

After being in Korea for six months, I requested to extend my time, so that I could command the company for at least a year. For me, staying there was a weighty requirement. I had left my wife and son in Baltimore with my mom and dad and now I would be gone even longer; not just for the year of the tour's initial length, but now that time would be extended because I was commanding a company.

The soldier signs up, but it is the family that serves. Joyce and I would make this decision together. In those days, we had two forms of communication: the U.S. postal system/military postal system and MARS, a military affiliate radio system, which included an operator in the middle.

For the initial three to four months, MARS calls are how Joyce and I talked while I was in Korea. On one call, I told her of having been offered company command, but to receive it, I would have to agree to extend for at least six months to have at least a year in command. We wondered: *What if she and KJ could come to Korea? It would not be an accompanied tour, and we would not be living together.* We discussed the command opportunity, the extension requirement and alternatives with the MARS operator listening to every word. Although our plan had been to leave the military after four years, one truth was clear: soldiering was growing on me, and I was enjoying it.

On an unaccompanied tour, families were not *command sponsored.* Command sponsorship provided travel, housing, medical, commissary, exchange and other privileges for families. In those days, the ability to be there and still have family member services provided was different than it is today.

In Seoul, near the Army Headquarters for 8th Army, there was an area where American families lived. I was stationed up on the DMZ where no families were allowed. This area was on the border with North Korea and the mission was patrolling. I thought that if I could find Joyce and KJ an apartment where the Americans were living, even though it wasn't an accompanied tour in those days, she and KJ could receive the benefits, minus housing, that would be there and used by families.

On one of my rare free Saturdays, I went to Seoul and was able to find an apartment complex where other Americans lived; it was safe and near the military installation. Joyce would have medical care, access to the commissary, and opportunities for her to engage in other ways.

After several MARS calls, our decision was made. I gave the go ahead and agreed to extend my tour to take command of Charlie Company, 1/17th Infantry (mechanized). Joyce and KJ would come to Korea; however, they would not be command sponsored.

The significant drawback of this in those days was we paid for travel and Joyce and KJ could not live in government housing. My place of duty, Camp Howze, was situated near the DMZ, and no families lived there. They could live in Seoul, where family members resided on accompanied tours. The combined Joint/Army major headquarters was located there, and service members were on two-plus year tours.

Joyce knew she wouldn't be living on a post, as we'd done at Fort Bragg. I told her, "You won't live in military housing, but you can live in housing where other American families reside, close to the military post."

With passports obtained, Joyce and KJ arrived in Korea in the summer of 1975. Their flight to Korea was an experience which included an unplanned overnight stay at the airport in Japan, with KJ tethered on a harness as they catnapped in the terminal. In an unfamiliar environment, she was very careful about making sure a restless KJ did not wander off as they stayed overnight in the terminal.

Once in Korea, they moved into that apartment. Other than not living in the military quarters, they could do everything else. Joyce and KJ had medical, commissary and exchange privileges. There were no issues with any of that. She was also able to continue to work as a just about full-time substitute teacher and did other things in the community. We found support for KJ in a Korean nanny that he loved and so his first overseas experience involved learning Korean culture. Almost 50 years later, he still remembers that.

Foxhole Lesson: Make the most out of situations when you are thrust into them, no bellyaching about it. Figure out how to make difficult situations work to your advantage.

Although Joyce and KJ spent almost a year in Seoul, Korea, our time together was limited. My company, Charlie Company, 1st Battalion, 17th Infantry (mechanized) was located at Camp Howze, near the demilitarized zone. Visits were confined to infrequent weekend visits. On rare occasions, I could get to Seoul for a weekend in our apartment. Or, from time-to-time Joyce and KJ could come on a bus and spend a weekend on my compound, staying in my room in the company commanders' hooch. When Joyce and KJ were able to visit my compound, all officers there were thrilled that they could connect with a 3-year-old; he garnered a lot of attention. They would sit him on the pool table with a pool cue in his hands. He knew he was playing pool, and they got an absolute kick out of it. With this arrangement, it was easy for the other commanders to get to know my family. They were the only family members there. Christmas found KJ overrun with gifts, and he got hives from all the excitement.

We became part of the community there and made friends—friends who today remember the times when KJ and Joyce came over. They saw KJ as a little guy and still remember those experiences. The military family dynamic is important, sometimes crucial. We have our nuclear family, we have our extended family with blood relatives, and we have a military family. That is real and it is significant and important too.

The command tour went well and although we didn't get a lot of time to spend together, Joyce and KJ's time went well too. On occasions I would make a trip to Seoul and she and KJ could come and spend a weekend with me on the DMZ, if my company wasn't on an active mission. Although we didn't spend a lot of time together, we knew the distance between us wasn't much, we were together. Joyce and KJ enjoyed their first overseas living experiences. Joyce got to teach in the American school and KJ started learning Korean, beginning his affinity for languages. It all worked just as it needed to work. And thankfully, every time the alert siren sounded, as it did their first night in the country, the alarms were practice alarms.

Combat Sports

In addition to combat missions, there were a lot of soldier-cohesion activities. General Henry "Hank" Emerson was the division commander; his nickname was "The Gun Fighter." He was the one who made the request to have senior lieutenants come to Korea, and he was still there in command when I arrived. He had these competitive games called "combat sports."

We soldiers lived with a lot of racial tension, and one of General Emerson's ways of relieving those tensions was having soldiers participate in these combat sports; regular sports like basketball, football, baseball and soccer, but played in combat style and behavior. So, more often than not, the games turned into free-for-alls! Yes, we had no minimum rules and organization, but that negative energy was released.

In these combat sports, company level officers had to participate. Captains and lieutenants participated right alongside enlisted soldiers. As a company commander playing combat basketball, you had to know what stead you were in with your soldiers when it came to whether they were setting you up for the other team to take you out! On the other hand, your soldiers could be taking care of you, keeping you squared away in these sports activities. Those combat sports were important as the Army dealt with issues of race relations, and it was an important part of the environment in those days. This happened throughout 1974-76.

The year in Korea, commanding C/1-7 Infantry went fast, and I would be saying goodbye to First Sergeant Alfred Hill after a great tour together. Hill was the company first sergeant, a career soldier, the senior enlisted soldier in the company. I relied upon his experience and wisdom in leading soldiers. He and I were a great team.

We, along with the lieutenants and noncommissioned officers in the company, were able to provide effective leadership that led to cohesiveness in our unit and conducted successful combat patrol missions in the DMZ; excelled at our gunnery training at nightmare range and other ranges; conducted force protection missions (Hot Apple) for our aviation and air defense units to the highest standard; and we reduced racial tension, replacing it with mutual respect and teamwork.

I left command as the first 2nd Infantry Division Company Commander without a report of survey in over two years. [A report of survey is an instrument for recording the circumstances concerning the loss, damage or destruction of Army property, and it happened often.] Soldiers under our command took accountability for their equipment; we followed Army supply and maintenance procedures.

Captain Kip Ward!

Having successfully commanded the company, at the 5.5-year mark of service, I had already been promoted. I was Captain William "Kip" Ward.

Now past my four-year commitment, Army life had been good: successful and informative lieutenant years in the 82nd Airborne Division followed by the successful command overseas of a mechanized infantry company with a real-world mission. A framework for the future was forming. My next plan included returning to Fort Benning to attend the infantry officers advanced course that had been scheduled 24 months before. Things were humming along for me at Fort Benning's Advanced Course.

Adding to that change, our second child was on the way—another indicator that leaving the Army was off the table in the summer of 1976. For some reason, news of another child gave me a boost of confidence, an overwhelming feeling of responsibility in the gut, and a sense of calm—all at the same time.

Kahni Jacinta Ward was born November 1976 in Georgia (at Fort Benning). When Joyce began going into labor, we took KJ next door to the Galloways, and it wasn't long after we arrived at the hospital that Joyce had given birth to Kahni. Our beautiful daughter, whose name was inspired by her big brother, was here.

One day the Army personnel assignments officer said, "You've done everything we've asked you to do well. How would you like to go to graduate school and then go teach at West Point?" That was an absolute shock because I had never considered anything like that.

> ***Foxhole Lesson: Day in and day out, do the things that you do at your very best so when opportunities present themselves, you are prepared to accept them.***

From Fort Benning to Penn State

While I served in Korea, an officer noticed my performance; he was a major, the battalion executive officer, and a West Point graduate. He knew West Point was seeking a more diverse faculty; they wanted a rotational professor to enhance the academic program for West Point cadets. Unknown to me at the time, then Major Waldo Freeman had written a letter of recommendation that I'd go there as an instructor. Obviously, the position required an advanced degree.

I had been an above average student. My grades in college were fine, and I passed the GRE (graduate record exam). With my undergraduate political science degree, I was selected to teach political science and public policy at West Point. Academic departments at West Point want their professors to attend certain schools for an advanced degree. Before assuming that position, I would need to complete a two-year graduate school program, and the Department of Social Sciences provided me with a list of grad schools.

One of the schools on West Point's list was Pennsylvania State University, State College. State College is a municipality in central Pennsylvania. It's home to Penn State University's main campus. I applied there and was accepted. It was close to Baltimore and the mid-Atlantic area where my family was located. Having been overseas for almost two years, we could spend time with family and our kids could be close to grandparents.

The Army sent me to graduate school at Penn State for two years—Yep! That was it—full pay and allowances, the whole nine yards. That was a pretty good deal. My sole mission was to earn a master's degree and to stay in top physical condition. I would take physical fitness tests regularly to make sure all of that was current.

Being close to Baltimore began paying off immediately. Joyce, KJ, and Kahni, our new daughter, would move to Pennsylvania while I attended graduate school there. Temporarily, KJ and Kahni stayed in Baltimore with my folks while Joyce and I went to State College to find a place to live. We moved to State College, Pennsylvania over the Christmas holidays.

Our arrival began another wonderful experience. While only a few years after leaving Morgan, a different trajectory from the "four years and out" plan was evolving. At every point there can be a change or an opportunity that alters or adjusts the planned journey. You enjoy what you are doing and the people around you. Increasingly, you enjoy the military community; it makes sense to you. You are doing what you enjoy doing. Importantly, your wife, four-year-old son and baby daughter are adjusting nicely and even thriving. They liked it. Joyce's impressions of the Army's unique opportunities were: "This works!"

Even though there were a couple of things along the way that were questioned—like the motor officer assignment, planning next to do an advanced course but instead being sent overseas on a short tour to Korea—those issues worked out in a way that led to a positive development. Now I was in graduate school for two years with a career trajectory to: 1) be a student and get a graduate degree; 2) go to West Point and teach cadets.

Precious Time with Family

The bonus of that course and the two years was that I had time to spend with my family and to get to know and grow with my two young children. I couldn't replace that. Watching those two grow up was wonderful. I could be with my son in his kindergarten class on one of those "What does your dad do?" days. KJ could say, "My dad's an Army soldier, a paratrooper." The kids enjoyed seeing a real parachute up close and hearing about jumping out of an airplane. They were mesmerized, and KJ was a hero. Taking KJ swimming and watching Kahni take her first steps were all unexpected, unforgettable treats. Those two years at Penn State and being a part of the community was foundational for me as a father.

Penn State was a large school, huge program and student body, with a big-time football program. As I take on graduate school life, I'm experiencing the scholastic and academic environment from a totally different point of view. The contrast between military and academic institutions is remarkable and informative, especially juxtaposed with five years of a military environment. As a married adult, seeing things from a different perspective and getting to know this community, my viewpoint and references expanded. This wasn't Baltimore, Fort Bragg, or Korea.

Experiences like managing married and military life, staying connected with family and civilian communities, and being current with scholastic, cultural, and professional dynamics are not what many of today's folks see as beneficial. Often, personnel are guided falsely with comments like, "If you do this or that you're going to ruin your military career." Well, for me it didn't. I know it was a unique time for me and things weren't going to be the same or easy, but life goes in cycles. What was unique and rewarding for me in 1978-79 isn't the recent narrative. I cherished every opportunity afforded by the time.

> ***Foxhole Lesson: At every point there can be a change or an opportunity that alters or adjusts the planned journey.***

These experiences added to the fabric of Kip Ward and the Ward family. My next iteration was understanding that wherever you go, you meet new people, and that place becomes home—an emphatic truth for military personnel who travel frequently. You have your birth home, but mostly home is where you are. You make that work. To this day, KJ and Kahni understand how to create and maintain a fulfilled life no matter where they happen to be. They know complaining about things is not productive but spending that time making the best out of where you are is beneficial—not only to you, but also to your community.

Something distinctive about an unfamiliar place always arises. People talk about their best tours or worst tours and assignments. Joyce says, "The tour is what you make it. The tour is about the people, not the location."

We've lived in some unique places, across the "lower 48" of the United States and from Alaska to Hawai'i, in Europe, Korea, Africa and the Middle East. The place is what you make it. During those days, State College Pennsylvania was a wonderful place to be.

> ***Foxhole Lesson: Understand that wherever you go and wherever you make that foxhole, you meet new people, and that place becomes home.***

From Penn State to West Point, NY

I completed my two years at Penn State in December of 1978

and earned a master's degree in political science. Leaving State College, Joyce and I did a permanent change of station move to West Point, New York. I was assigned to the staff and faculty at the United States Military Academy in the Department of Social Sciences as an assistant professor. This began a very gratifying tour from many perspectives, family and professional.

Reflecting on my first five years in the Army, while assigned to Fort Bragg with the 3rd Battalion 325th Infantry 82nd Airborne Division, was a high tempo operational atmosphere, where a lot of training was going on. We would engage in deployments to places like Panama, Puerto Rico and sitting on the *green ramp* for days during the 1973 Yom Kippur War. Simply not knowing what would come. We didn't go, but our mission was to be ready to go—*wheels-up* in 18 hours after alert. (*Wheels up* is military and air traffic controller lingo that communicates a flight's progression—departing an area.)

Then going to Korea, commanding my mechanized infantry company, on mission alert, conducting combat patrols and manning observation posts along the DMZ was intense. My first five years in the military saw a lot going on with significant family separations. Returning from Korea, attending the advanced course at Fort Benning, and then selected to go to graduate school for two years, followed by an assistant professor tour at West Point was a stark change for me and our family. That built a level of stability that is uncommon but so richly rewarding.

We formed a new family of friends at West Point. There were permanent faculty members and rotating faculty like me, along with the staff. Although they have permanent professors, West Point brings in the rotating faculty. This segment of the teaching staff of professors brings their recent experiences in the military to the cadets. That becomes a part of the cadet educational process. Cadets hear and learn from officers who bring recent military experiences to their classrooms. This provides a healthy mix of officers who are permanent professors as well as those who come in for a three-year teaching tour and then back to active duty in the regular Army.

Foxhole Lesson: Often, unexplainable situations are what prepares you for future surprises.

Our military family was continually growing. For most, leaving one place doesn't mean forgetting the people. In the military, we always met new people at the next duty station.

For me, Joyce and the kids, West Point was that type of an experience. When my kids were in elementary school, doing their extracurricular activities, I had schedule flexibility. I could take them to school. Joyce picked them up in the afternoon. I was able to coach some of their sporting teams, drop them off at school, and be with them when I couldn't be there before.

As a West Point rotating professor, I was able to bring my experience as an infantry lieutenant in the 82nd Airborne Division and as a mechanized infantry company commander in the 2nd Division, operating along the Korean DMZ. I brought all of that into the teaching of political science, public policy, and American politics. I integrated being a soldier, an officer, platoon leader, and company commander into their West Point cadet experience.

My time at West Point included extracurricular programs and cadet activities. These are a part of the educational setting at any college or university across America. My department, social sciences (SOSH), participated in staff and faculty intramural sports, notably basketball and softball.

Socially Joyce and I sponsored cadets. I was also the assistant officer in charge of the cadet gospel choir. I traveled with the choir up and down the eastern seaboard; they performed for churches and at concerts as ambassadors for West Point. These performances were instrumental in exposing the U.S. Military Academy to populations and communities who may not be aware of the opportunity for a college education available through the Academy.

Many of my cadets I taught later worked with me. I was able to promote them and prepare them. I witnessed many of these cadets serve with distinction on active duty; like Lieutenant General Bill Mayville becoming the Director of the Joint Staff and Lieutenant General Nadja West becoming the Surgeon General of the Army.

I was leaving West Point where I had been a professor and able to spend time with the kids and Joyce. I could do more with the kids and be with them. I even learned how to plait my five-year-old daughter's hair. Joyce would do the prep, but I would do the final touches and what not. Learning that was a treat for me—I later learned not so much a treat for Kahni. I distinctly remember my baby telling her mother her preference about who plaited her hair: "Dad does these little tight plaits. He makes them so tight and skinny."

I was trying to be so precise and have everything tight—that might be military training residue. Her complaint underlined the fact that military tightness isn't what one goes for when styling a girl's hair. But we worked through all of that, and it was a good time for me being able to spend time with KJ and Kahni and be part of their activities. They were 9 and 5, and for the first time, I could be right there with them. Given how things moved on in my career, it was one of the last times as well. I cherish and am thankful for those West Point years.

At West Point, Joyce had taken a professional job as a government civil service employee. She worked in the Civilian Personnel Office (aligned with her undergraduate business degree) in various personnel positions and conducting various actions. We often lamented over one of her actions in overseeing a reduction-in-force for civilian workers and were able to share things like that. We had a rich life, and it wasn't because we had more or less than anybody else. We took advantage of where we were, made where we were home and caused ourselves to be integrated into the community and that caused the experience to be a great experience. All these things enriched and enhanced the lifestyle that we had, and it all worked.

I met many officers and friends, service members who till this day are the dearest of friends that we have. It is because of how we were able to interact and form bonds there with respect to our social activities and supporting each other and supporting cadets. Many of the cadets that I taught keep in touch with us today. My former students have retired, but I encounter them regularly.

We met all those people back in the early 80s that we are still so proud of. We continue to live like that. You don't know what it is that causes you to go someplace other than some divine activity; but the point is that when you get there don't belly ache; you take advantage of it, make the best of it and you'll find yourself much better off because of it. That has been the case for every experience that I've had; from the time on Falls Rd. in Baltimore in that little Black community to my dad building our home; going through elementary, junior and high school, going through college, graduating, meeting my wife and getting married. This type of thing continued to happen over the course of my military career.

> ***Foxhole Lesson: Sometimes assignments may not make sense and—other than some divine activity—you won't know why or what caused you to be there. Once you are there, don't belly ache; take advantage of it, make the best of it.***

Glad to Be Dad

Having this precious time in my children's lives made me think of the times when I was absent. In the 82nd Airborne Division and in the 2nd Infantry Division I wasn't around a whole lot with my family. There were training exercises and assignments that took me to various places like Panama and Puerto Rico. During much of my career, family time was irregular because my duties were not within a 9-to-5, Monday-Friday schedule.

I became a member of the first elected school board council on West Point. I taught political science and public policy, so running for a somewhat political position was interesting. I mounted a campaign and became a member of the first elected school board at West Point. Another instructor in the Department of Social Sciences was my campaign manager, and I had my stump speech: "WARD: Willing, Able, Ready, Dedicated! Vote for Ward for the School Board!" Had I not been in those places, I never would have experienced or even thought about these kinds of opportunities.

Joyce and I packed up KJ and Kahni and left West Point in 1982 after I'd been selected to go to the United States Army Command and General Staff College at Fort Leavenworth, Kansas, an intermediate level school that the Army sends a select group of its officers to. Selection is based on demonstrated duty performance and potential. I was ready for a year of study prior to the next assignment (whatever it might be)!

Notes:

- ***Yom Kippur War: On October 6, 1973, an Arab coalition of Egyptian and Syrian forces launched a surprise attack on Israel on Yom Kippur—the Jewish holy day of atonement.***
- ***A brigade typically consists of three or more battalions and anywhere from 3,000 to 5,000 soldiers. A colonel is generally in command. The brigade commander has a self-contained headquarters and staff.***
- ***The Korean Demilitarized Zone (DMZ) served as a buffer zone and demarcation line between North and South Korea.***

CHAPTER 3

More Foxhole Lessons

With our caravan of two cars and a trailer, we took off on our tandem journey from West Point, New York to Fort Leavenworth, Kansas. We hooked the trailer my father had built to the car with KJ and myself; Joyce and Kahni were in the other car.

On that drive, we stopped in Ohio to visit Joyce's brother, Levi Lewis, Jr., who was married and living in Dayton. We were happy that KJ and Kahni could connect with family members, really see them, get to know them more, and spend time with them.

My assignment was to attend the U.S. Army Command and General Staff College (CGSC) full time, Monday through Friday, with no weekend duties, and more time for family. Other than not working with cadets, this was like West Point.

Fort Leavenworth was a permanent change of station move; that meant moving all our household goods. After so many moves, we were familiar with managing the process. We set up our household and lived there for one year while I went to this intermediate level staff college.

KJ and Kahni attended school on post. Joyce continued her professional career path as a human resources specialist. The position she landed found her working in the disciplinary barracks at Fort Leavenworth, a military prison. At West Point, she managed programs impacting career civil servants; at Fort Leavenworth, she worked with prisoners' personnel records. Flexibility matters and this was a case where spouses were limited in the opportunities available for them to continue their careers. The Army and other U.S. Defense Department Services have done a lot to address this.

From a professional point of view, we were able to continue the recent trend of having a flexible, not very exhaustive, and rewarding schedule. My workday of classes began at 8 a.m. most days and ended around 4 p.m. Some days were full; other days I only had one or two classes, it depended on the day. The predictable routine gave me flexibility to do things at the house and with the children in their activities. We really enjoyed and treasured our time at Fort Leavenworth. The motto there was, "This is the best year of your life," and it was a wonderful assignment.

At Fort Leavenworth, Joyce and I reconnected with several military friends, former Morgan classmates, and met new friends also selected to attend CGSC. Meeting up with these friends from prior assignments at Forts Benning and Bragg, West Point and Korea made for a great reunion.

We quickly understood and embraced the dynamic of having a blood family and a military family. In some respects, it is just like the old neighborhood. Military family relationships are just as strong as family ties. These are folks who would do anything for you. We cared for one another in some very deep ways. We loved them, appreciated them, and enjoyed being with them.

Fort Leavenworth, although an academic environment, was sort of a re-greening for me. I had been to graduate school for two years at Penn State, followed by an assistant professor position in an academic department at the U.S. Military Academy at West Point for three and a half years. In my mind: *This is my next step in getting back to the "real" Army and duty with troops.*

Family Decision

In the spring at CGSC, I became Major Kip Ward. The class was composed of different year groups—this means the years that various attendees would have been commissioned. At this 1983 graduation, I had been commissioned 12 years earlier (1971). Others may have been commissioned in 1968, 1969 and 1970, thus one to three years senior to me. And then there were those who were commissioned as late as 1973. So, there were officers who had been commissioned over a period of five or six years.

I had arrived at the CGSC as a promotable captain and was promoted to major before graduating. Following CGSC, I was looking forward to an assignment back with troops in an infantry unit.

At the 12-year mark, as Major Kip Ward, the die was cast, and Joyce was integral in that. No decision like that would have been made without her full and positive concurrence. We made the decision to be around for at least 20 years. It was an investment that made sense for us. Attending law school was now on the back burner, and it was clear I was probably going to be a career soldier. By our actions, we were committing to stay in the Army.

At that time if you stayed in the military for 20 years, you could retire; although retirement would not provide a decent living wage for a family of four. You were, however, young enough to start a second career. This was something that many soldiers did and still do.

KJ and Kahni played a role in that decision as well. We wanted to make sure they were also OK with continuing our journey in the Army. At 10 and 6 years of age, in their short time on earth, theirs was a life that had become familiar and comfortable. The move wasn't something that posed a problem for the kids. They embraced it saying, "Hey, we're military brats and proud of it." There was no competition between our blood relatives and our military families in their minds. Our time at Penn State and West Point reinforced and enhanced their relationships with grandparents, cousins, uncles and aunts.

Thinking about requirements and tasks I would have to achieve as a major, my next significant objective would be to command a battalion as a lieutenant colonel. We called it being *branch qualified*. The terminology is different today, but it simply meant being qualified to assume positions in your branch at the next higher grade.

From an assignment point of view, the older officers, who had been promoted to major two or three years earlier, would be receiving priority for branch-qualifying assignments. They needed the assignments before their next promotion window opened.

In my case, I had been away from the active Army for over five years; two years in graduate school and three and a half years at West Point. So, although I was a new major, I needed to get back to a troop assignment, preferably at the battalion/brigade level. I was an infantryman, and I wanted to get back to an infantry outfit to be competitive for follow-on assignments.

To be promoted to lieutenant colonel and command an infantry battalion, I needed to excel in a branch-qualifying assignment and the branch-qualifying positions for infantry battalion command. As an infantry major, I needed to perform successfully as a battalion or brigade S3 (operations officer) or XO (executive officer) to be considered for battalion command, a lieutenant colonel position. With my time away from troops, I looked forward to being assigned to an infantry battalion or brigade as the operations officer or the executive officer.

A Friend Who Sees Me

During the second half of the CGSC course, the Department of the Army Personnel Center sent assignment officers representing all the branches to Fort Leavenworth to review and tentatively make those assignments. This decision would affect my going to an infantry battalion or brigade. I knew that not completing those assignments would not bode well for my future. But that is not what was presented to me. I asked the branch representative why I was not going to one of these infantry jobs.

His response sounded like a song and dance, hemming and hawing, "Ward, you will not be going to an infantry unit. You're a junior guy and have time to catch up." That is a comment often heard by minority officers when the "system" denies them opportunities that others in the same situation receive.

Here we go again, I thought, *another motor officer type scenario that minority officers face at a higher incidence than our majority counterparts.* The rationale given was: "We need to give the more senior majors those jobs, as their timeline to promotion and command selection is shorter."

I saw right through his obscene reasoning as I discovered there were other officers, who didn't look like me and were my junior, being assigned to infantry units upon graduation. The old system was in play as to why I wasn't getting assignments that could help me move forward, yet others were. That didn't sit well with me, and I expressly displayed my dissatisfaction with the decision.

During the infantry assignment interview process, one of my good friends, also an infantryman, looked over and saw me rise, pointing my finger and gesturing. It truly upset me for this guy to look at me, saying those things when I knew others were being treated differently. Clearly, a standard was applied to me that was different for my majority counterparts. Therefore, I knew that I had to make something happen for myself.

The friend who saw me get upset was Charles "Chuck" Ware. Seeing how it impacted me, he called Joyce and suggested she take me somewhere for me to calm down a little bit: "You and Kip just need to go out somewhere for dinner and we'll take care of the kids."

Chuck and Kate Ware had a boy and girl about the same age as KJ and Kahni. We had met several years earlier while attending the infantry officers advanced course at Fort Benning. (Chuck went on to command an infantry battalion of the 24th Infantry Division in the first Gulf War. An outstanding commander, his was the lead Bradley battalion spearheading the drive to Baghdad.)

In the end, I was told that I was not going to an infantry unit; instead, I was going to an assignment as a foreign area officer (FAO), an assignment for persons working in embassies or with foreign militaries. My graduate degree in political science for teaching at West Point, along with the international flavor of the degree led to assignment officers grouping me in that category and camp. FAOs had limited opportunities for assignment and no command billets. (The type of treatment FAOs received led me to have a great appreciation and become a staunch advocate for these soldiers and that appreciation increased as I gained more seniority in the Army. They are among the unsung heroes for our nation.)

My initial assignment would be working overseas with the Saudi Arabian National Guard. I realized that a two-year assignment there would reduce the probability of ever being selected for battalion command. I would not have the opportunity to become an infantry executive officer or operations officer. FAOs were not well respected in the military; therefore, I went back to the Army Personnel Command assignment officer and told him I needed to get back to the infantry because at that point I'd been away from the infantry for five years and wanted to be eligible to command an infantry battalion.

Send Me to Germany

I talked to him about overseas assignments in Germany, where the Iron Curtain was still up. This was before the end of the Cold War. We had a lot of forces stationed overseas in Germany. The assignment officer said, "That just isn't going to happen."

I wasn't willing to accept that. There had to be something. There had to be a way that I could get to an infantry unit, even if it was a circuitous route.

As I considered options, it was not lost on me that I had never been to an infantry outfit where there wasn't a logistician. Logisticians dealing in supply, transportation and maintenance were needed to make operations happen. As a lieutenant, I had been a battalion motor officer, and I knew that battalions and brigades had logistics officers. Infantry units had logisticians. I thought: *If I get to an environment where infantry units are around me, I will find a way to get assigned to one.*

To get to an infantry unit meant being assigned to a non-infantry unit but in a location that included infantry units. If I could work in one of these non-infantry jobs near infantry units, I'd eventually get to an infantry unit and the branch-qualifying job of S3 and/or XO. I reasoned: *OK. I can't be assigned as an infantryman, but maybe I can get to the environment as a logistician and then eventually get myself into an infantry battalion.*

I went back to the Department of the Army and inquired about opportunities for majors who were CGSC graduates. That level of schooling made us a priority for certain assignments. They said, "Well, you can go to Germany, but you have to be a logistician."

I quickly responded, "Aha! OK. Send me as a logistician to Germany."

I knew that if I got to Germany, I could find an infantry unit and sooner or later be assigned there. At that time the Cold War dictated that many infantry units were in the U.S. Army in Europe.

In Germany, the United States had two corps: the United States 5th Corps and the United States 7th Corps. Both had two divisions. And there were other tactical units associated with the U.S. Army in Europe—from armored cavalry regiments (all combat units) to field artillery units. That meant there were a lot of opportunities in Germany. I felt that if I could get to Germany, we could make this work.

The Army granted my request to an MOS (Military Occupational Specialties) school. In this case it was 92 Quartermaster, logistics; 92 is the code for the quartermaster branch (supply). The school trains quartermaster personnel, officers and enlisted, and is located at Fort Lee, Virginia. I could go to that school to get qualified as a logistician and then get assigned to Germany. However, it wouldn't be as an infantryman; I would be a logistician, but it would be in Germany where other possibilities lay. Thus, I would have three specialties: primary branch infantry (11), alternate of foreign area officer (FAO/48); and lastly quartermaster (92) logistics.

My challenge had been to figure out another way to accomplish my goals. When the Army placed unexpected obstacles in my path, that became another example of how a complicated, unpredictable circumstance appeared and worked itself out in the end.

Many instances like that happened to and for me: like that first motor officer's course and being assigned as a battalion motor officer. At the time, I just looked at it as my lot and what I would do; but later, it paid off. Once again something that started out as less than optimal ended up in us making the best of it, and the good Lord took care of the rest.

> ***Foxhole Lesson: If there doesn't seem to be a straight line between where you are and where you want to be, draw yourself a line that gets you to a place that puts you in a position to get to where you want to be.***

The summer of 1983 at Fort Lee was wonderful for us. I was getting branch qualified as a logistician because that was the only way I could ultimately get back to an infantry unit. The dividend for our family meant we were able to spend time with Joyce's family. At the time her grandfather, Rev. William Elvis Hopson, was still living, so our kids were able to spend time with another great grandfather. Although Joyce's father had passed away when she was eight years old, we were all blessed for the time we had with Joyce's mother, Mamie Hopson Lewis. What a wonderful lady. We look at it now and appreciate how it was a blessing for us. We knew it was a blessing, but how it all began had been unexpected and different.

We had a great two and a half months in Virginia, and in August of 1983, Joyce and the kids and I packed up from our temporary duty in Fort Lee and made a permanent change of station move (PCS) to Herzo Base to live in a town we'd never heard of: Herzogenaurach, Germany.

The other part of it was that I was an infantryman and although I was a logistician, I was going to a field artillery unit in Germany. Hello 7th Corps Artillery, here we go! I was a brand-new logistician—just finished a two-month school teaching me how to be a logistician. I knew about maintenance because I had been a battalion motor officer in the 82nd. I had commanded a heavy (mechanized) infantry company in Korea. I was ready to be the logistics officer in this corps level heavy artillery brigade, as the Brigade S-4.

This heavy artillery brigade of four battalions in Germany with a real-world mission during the Cold War and before the fall of the Iron Curtain in Europe was the force artillery headquarters for the 2d Armored Cavalry Regiment (ACR). (The 210th Field Artillery Brigade and its subordinate battalions provided long range indirect fire support for 2ACR.) The nuclear capable brigade consisted of two 8-inch, self-propelled artillery battalions, a 155 mm self-propelled artillery battalion, and a Lance rocket artillery battalion. The Lord had prepared me for that assignment.

In earlier assignments, I had performed tasks, developed and acquired unique skills. Sometimes I had questions about the process and wondered about the purpose. And now here was my answer! At Herzo Base, I could see how that training prepared me with the tools I needed to be successful in this job. As I'd learned in my ROTC training at Morgan State, whatever job you have, be the very best at it and leave it better than you found it.

Those Unique German Homes

Going overseas on an assignment typically meant military personnel and families would be sponsored by Army personnel already living there. The sponsor knew our family composition and met us at the airport in Frankfurt for the drive down to Herzo Base, right outside of Nuremberg, Germany.

Now we were in Germany, but we hadn't secured a house yet. Hotel arrangements were made for us until we could find a place to live. We wound up living near Herzogenaurach, a small community right outside of Herzo Base. The post housing office helped us find a nice German home about five minutes away from the installation. It was called "living on the economy." We had German civilian neighbors as well as other military families nearby. We were surprised to see the Adidas Shoe Factory located in that small town.

Small villages bordered the military installation, most were former German Kasernes (installations). Our home was indeed a German house. Imagine moving into a house where you literally bring your own kitchen sink and refrigerator. It was that kind of environment with the German homes, and we found this same thing in off-post housing. When the Germans moved out, everything was taken. In many instances even light fixtures moved with the former tenants and had to be reinstalled. This was another remarkable experience as a total family, now living on the German economy in Herzo Base, Germany.

I quickly became engrossed in my duties as the Brigade Logistics Officer for the 210th Field Artillery Brigade. As the force artillery headquarters for the 2nd Armored Cavalry Regiment, our active mission was to protect and prepare for an invasion from the Soviet Union. That was a real-world mission, and we regularly practiced moving to our general defensive positions (GDP). We regularly conducted gunnery exercises at the Grafenwoehr Training Area. It was impossible to travel on the German highways, the Autobahn, and not see large U.S. military convoys going and coming from training areas.

In addition to being the senior African American officer on the brigade staff, I was the only infantryman and logistician. Everyone seemed to question how an infantryman got assigned to this artillery outfit. I often found myself talking about the assignment process out of CGSC that I had to overcome. The linkage about being an S4 came into play, and I made a point of letting folks know that I eventually wanted to be reassigned to an infantry unit. I got remarks like: "So you're the logistician. So that's kind of a branch immaterial. Logisticians are everywhere." And that had been my thinking. The nuance was that I would be assigned as a logistician, but in an infantry unit. As it happened, however, my assignment was in an artillery unit. Consequently, I would still have to get to an infantry outfit for my branch-qualifying assignment as a major.

Infantry officers don't get assigned to artillery units as the S3 (operations officer) or the executive officer. So even though I was there in Germany, I still had the challenge of getting to an infantry outfit. At least I knew that infantry formations were around me. The Third Infantry Division and the First Armored Division with their infantry units were a part of the U.S. Army, Europe. But this wasn't the time to make getting to those units a priority. At this point, my mission was to be the very best S4 that I could be in this artillery unit.

This was a nuclear delivery capable artillery brigade with:

- Two heavy 8-inch, self-propelled artillery battalions,
- One medium 155-millimeter self-propelled artillery battalion, and
- One Lance rocket nuclear capable artillery battalion.

These were heavy armored mechanized outfits, with numerous tracked and wheeled vehicles.

As a brigade logistics officer, I had to be nuclear surety certified, because the Lance missiles were nuclear capable delivery weapons systems. As the S4, I would be responsible for the supply and maintenance of the entire brigade, including its ammunition resupply requirements and the basic load of ammunition used to go to war. Those responsibilities included operating the ammunition supply point (ASP) where the munitions were stored.

My learning curve continued to expand. It included the

maintenance and readiness of vehicles and the weapons systems, making sure the equipment in the units was operational. Having been a motor officer in the 82nd Airborne Division and company commander of a mechanized infantry company in the 2nd Infantry Division prepared me well for the job. My abilities had to be at a high level because we always had to be ready to go and expectations were high. I spent a lot of time working with the units to ensure that their equipment was combat ready and maintained, ammunition was stored properly, and the troops had everything they needed for doing their jobs—the tasks that logisticians do. This included the care and feeding of soldiers in the field and in garrison.

I had been the mess officer as an additional duty while a company executive officer as a lieutenant in the 3rd/325th Airborne Infantry, 82nd Airborne Division. This assignment took that to another level as I was now responsible for the dining facilities in the brigade, including renovations and design.

Food service was a part of my responsibilities, along with my team of soldiers. I had a great team who did outstanding work, including a warrant officer who was probably one of the best in Germany when it came to designing and managing dining facilities. Troops were fed right and efficiently, and they enjoyed tasty and wholesome meals. My team and I had oversight responsibility for managing those systems.

Fortunately, the Army's system of experts residing in its warrant officers includes food service warrant officers, who are true experts in these matters. Chief Warrant Officer Jose Rodriguez worked with me in the Brigade S4, and he was a master. I learned a lot about food service operations from him. With that knowledge later as a higher-level commander, at the battalion, brigade, division and higher levels, I was acutely aware of how to assess my unit's dining facilities to ensure they were always above standard. This was important because providing quality meals for soldiers is inherent in their care and an important aspect of mission readiness.

During this time in artillery units, the Army was fielding a field artillery control system called TAC-FIRE. It was a tactical fire system used to automate the fire control, the call for fire, and the delivery of artillery munitions on target. Helping to field TAC-FIRE amongst these units was also part of my responsibility. Although not a field artillery expert, I was dedicated to doing my best in making this fielding a success. I did, and it was. I had not given up on moving to an infantry unit, but in the meantime, I would perform at my very best.

I would travel all over the VII Corps area in southern Germany developing relationships for enhancing the readiness of the brigade. I looked for additional sources of repair parts, working with adjacent units and with our support command headquarters in Stuttgart, where the main logistics sites were. It involved getting and procuring materials and supplies to keep my brigade, the 210th Field Artillery Brigade, as ready as it could be. We performed with an operational ready rate routinely exceeding 92%, often cited as the best in the whole of VII Corps. For most combat ready units being at and above 80% is the standard.

Achieving High Standards of Unit Readiness

After the first year, when the time came for my officer efficiency report (OER), I was confident that my S4 team and I had excelled and that we would receive a rating that reflected that excellence. In VII Corps, in addition to my artillery brigade, there were two other artillery brigades and two Army divisions: the 3rd Infantry Division, the 1st Armored Division; a Combat Aviation Brigade, the 2nd Armored Cavalry Regiment, the Corps Support Command, among other subordinate formations. We were all competing for supplies and equipment in our quest to maintain a high readiness rate.

I was determined that the brigade would maintain the highest readiness rate; this included coordinating and establishing high priorities and building relationships with those who were responsible for helping our brigade stay ready, especially the technicians and specialists, warrant officers and NCOs at the Corps Support Command. As a result, we were doing some things that were innovative and significant in maintaining our readiness.

Our success was obvious and commanders of other units, thinking their outfits should have higher maintenance readiness rates, were perplexed at how we were able to continue to perform so well. We were doing a lot of coordinating and going the extra mile (literally, in some cases) to find what we needed; and the team of maintenance personnel across the brigade was with me in that effort.

Foxhole Lesson: Work extra, do more than is expected, and stay prepared.

We forged a team that just hummed along making things happen in some very real, meaningful and exemplary ways. We did food service, maintenance, supply management, repair parts management, and ammunition accountability better than most and kept artillery systems and vehicles in a high state of maintenance readiness. We even designed and outfitted dining facilities.

Some of our motor pools were covered with gravel. Gravel motor pools in Germany meant soldiers working and maintaining in mud, snow, etc. Soldiers working on tracked vehicles, laying in mud and rocks and gravel wasn't ideal. We could do better than that, and we did.

We made improvements in those areas as well, thus leading to the improved readiness. Yes, the mission is first, but taking care of people is essential to accomplishing the mission.

Later in 1992, as Brigade Commander of the 2d Brigade 10th Mountain Division, I made our motto: "Mission First, People Always." Almost 20 years later, Army Chief of Staff James C. McConville adopted that motto at Army level.

Considering all of this, when it came time to receive my efficiency report, I had definite expectations of a top-block report. By every measure we were exceptional. I met with my brigade commander, but the report from him did not reflect that level of performance. I looked at him questioningly.

He said, "You know Kip, I have to save my very top blocks for the artillery officers I rate and you're an infantryman. You're not one of the battalions S3 operations officers or executive officers, but you'll be OK."

This was a blow, but a scenario I had faced before. Yeah, I was good. Real good. But regardless, I would be behind counterparts. Saying "You'll be OK" or "You can wait until the next time" is a familiar refrain heard by minority officers. Many don't recover from those slights.

I felt that it just wasn't right. While I served this brigade, systems and processes had been enhanced, and my team made remarkable accomplishments, yet my team and I didn't get the report we deserved, and others knew it as well. Peeling back the skin of this onion—denied a higher rating—revealed the excuse: *You're not a field artilleryman.* But that was a shallow explanation for what I certainly knew was reality: *Who I was and what I looked like made me "expendable."* I was competing with more "favored" peers despite the excellence of my work, and I did not receive the rating that I and others thought I deserved.

I was not the first Black officer to receive this type of treatment. In addition to the personal aspects of the sub-par rating, it also caused me to be sensitive to not disadvantaging someone in a low-density job in my unit. This lesson was both personal and communal because it impacted others influenced by my commitment for always doing the right thing when it came to an officer's efficiency report (OER).

[Later in my career when a minority battalion commander in the division I served in faced a similar slight, I brought the situation directly to the attention of the division commander. The officer had received a good OER, but because he commanded a low-density battalion, he was not cited as being the best commander, regardless of branch. Thankfully the commander acknowledged the distinction and changed the report. These slights may be subconscious occurrences at the time; however, they happen disproportionately to minority officers, as I've witnessed first and second-hand.]

For my part, there wasn't a whole lot I could do about the report. The knocks were subliminal. On the surface, the report wasn't bad, it simply did not reflect the level of excellence that we had achieved, nor the degree that would ensure me remaining extremely competitive going forward.

It would be the next brigade commander who pushed for my reassignment and (after a year with him and achieving the same high standards of unit readiness) provided an OER that reflected the excellence of our work. That was Colonel Charles Hood. He was now my boss. I was elated! Colonel Hood knew that I was a major and an infantryman in his artillery outfit.

He asked, "What are you doing here?"

I responded, "I'm your S4!" Having him as my commander would be a wonderful experience for us.

I was now in a situation like the one I experienced in Command and General Staff College. If you want to command an infantry battalion, you need to be in an infantry outfit. Colonel Hood knew I couldn't stay there very long. I had no thoughts about special treatment, but I knew here was someone who would be endorsing and supporting my efforts to do the things that we all knew were important for me to do to be successful. Even with the threat to the brigades' excellent readiness posture, Colonel Hood took the lead in convincing his boss that releasing me was important for my career. In Chuck Hood, I had an advocate.

And there would be others to inspire me. Through Colonel Hood I was introduced to Colonel Johnnie Wilson who commanded the Division Support Command in the 1st Armored Division. Other officers included: the direct support maintenance battalion commander, providing the direct support maintenance for our brigade, Jim Monroe, a promotable lieutenant colonel; and commanding one of the major U.S. training centers in Germany at Hohenfels was Infantry Colonel Don Scott. These men became my mentors.

Chuck and Elaine Hood and Joyce and I were close friends. The Hoods would have gatherings with other officer families, and they invited Joyce and me to come. We were the junior people around these senior officers. It was quite an experience to interact with them. Although unspoken, the consensus from them was, "We've got to get Kip out of here and back to an infantry outfit where he can be competitive for infantry battalion command."

We spent that second year there working with the Hoods. It went well. The role that we played in the readiness of the artillery brigade led to a very high level of success and achievement.

During that time, the Field Artillery Association presented me with the St. Barbara's Award, named for the patron saint of the artillery, St. Barbara. The award is given to those who enhance the profession of the Field of Artillery. Field artillerymen are eligible and typically receive this coveted award for contributions to the field artillery.

On occasion honorary awards are presented to non-field artillery persons. Unbeknownst to me at one of the events, I was presented with the St. Barbara Award based on my substantial contributions. It was not an honorary award. On top of that—for spouses who provide extraordinary support to the readiness and morale and care of families—the artillery presents the Molly Pitcher Award. To our surprise, they honored Joyce with the Molly Pitcher Award also!

Here Joyce and I are—in this artillery unit, with these two awards that are presented to members of their profession, and we received them. The presenters emphasized more than once that except for honorary awards, it is rare for someone outside of the field artillery to receive these awards. They emphasized the point exclaiming, "No! You and Joyce received these awards because you earned it; you've supported and contributed just as others who are a part of the profession and get recognized."

That both of us were recognized for the roles we played in enhancing the morale, welfare and readiness of that brigade—both in its war fighting mission and in caring for families—was deep and meaningful. *We bloomed where we were planted.*

Extra special and so rewarding was that Joyce was right there, a part of the environment, fully integrated and doing the same kinds of things. My boss, Colonel Hood, was someone I knew would support my efforts to move forward and beyond to the next steps.

After two years in the field artillery brigade, I still was feeling the need to get back to infantry and do well in a branch-qualifying job to be competitive for command of an infantry battalion. Chuck Hood said, "That's exactly right, and we're going to figure out how to make that happen."

Colonel Hood was instrumental and supportive in working with the divisions that were part of the 7th Corps, so that I could transfer from 210th Field Artillery Brigade to a divisional unit in the 3rd Infantry Division.

For us, Europe was a good thing—it worked. Although things don't go the way you envision, you take what you get, make it work and move forward. This doesn't mean that you don't advocate for yourself as fiercely as you advocate for others. It means that when all is said and done, when the dust settles, you pour your all into where you are. I received an assignment to the 3rd Infantry Division, but before getting to an infantry outfit, I had to wait for others who were in line ahead of me. People who had been on the division staff were in line waiting to go to a brigade or battalion as an S3 or XO. The idea was: Before you get yours, you must be there for some time and be known. In some regard I thought to myself, "Here we go again." My initial assignment to the 3d Infantry Division in Aschaffenburg, Germany was to a totally unexpected position.

Foxhole Lesson: Although things don't go the way you want them to go, you take what you get, make it work and move forward.

It's now the summer of 1985, two years out of CGSC and two years in grade as a major. I was in an infantry division, but I would be working as a community executive officer. What will this be about? It turned out to be one of the best jobs I could have ever had! This position exposed me to the community and installation support side of the Army and brought opportunities to work for the quality of life, well-being and welfare of soldiers, families and civilians. It revealed behind-the-scenes activities and responsibilities—tasks that aren't highly visible or glamorous but are essential to mission readiness and whose neglect can compromise it.

As the Aschaffenburg community executive officer (XO), I worked for the Deputy Community Commander, Lieutenant Colonel Dick Benjamin. My responsibilities included oversight of community activities:

- Commissary operations and post exchange to the Department of Public Works
- Department of Engineering and Housing
- Civilian Personnel Office
- Liaison to the on-post schools and more.

So, before my assignment to the infantry battalion—1st Battalion (Mechanized), 7th Infantry (Cottonbalers)—I learned all about the community support activities that soldiers and families depended on for their quality of life and wellbeing; that included behind-the-scenes services:

- Schools
- Medical facilities
- Post exchanges

- Commissaries
- Engineering and housing
- Civilian personnel
- Childcare
- Army Community Services
- Training areas and firing ranges, etc.

Most importantly, I firmly understood how each of these activities was integral to the combat readiness of the units.

The contract was that I'd serve as the community executive officer for a year and then move to a Bradley (mechanized infantry) battalion. I said, "OK," despite feeling that I was once again going into a job that wasn't aligned with the path I had envisioned for myself. But I was determined that I was going to do my best, be the best that I could be. I would learn the position and make it work—and we did! We left Herzo base near Nuremberg and moved to Aschaffenburg, close to Frankfurt.

I began the job as community XO and my boss, Lieutenant Colonel Dick Benjamin, put a lot on my shoulders. I eagerly tackled it all. In this position, I was exposed to a whole new segment of installation support activities that I knew were there but knew nothing about. I soon discovered how critically important successful management and operation of these activities was to combat readiness. This included the management and operation of training and gunnery ranges.

Beyond the community job, my intent was to get to an infantry battalion as an S3 or XO. That is what I needed to do to become eligible to command an infantry battalion. The goal of being branch qualified was always before me. Now in my third year being a major, I was not branch qualified because I hadn't gotten an XO or S3 job in an infantry outfit. During those years I had filled positions that were not my first choice, but those positions had set me up for performances down the road, preparing me and solidifying my success as I assumed higher positions of responsibility.

That community XO job taught me so much about how the Army runs; those intricate systems that support the Army and that support the warfighter. Indeed, I drew on those experiences later. I was now in command of an organization that relied on those things to function and be successful. I spent a year as a community XO and finally got to a battalion. I was the Executive Officer (XO) of the 1st Battalion (Mechanized), 7th Infantry (Cottonbalers), in the 3rd Infantry Division in Aschaffenburg, Germany. That was a Bradley Infantry Fighting Vehicle Battalion and one of the first Bradley units fielded in the Army.

Bradley vehicles were designed to transport infantry or scouts with armor protection, while providing covering fire to suppress enemy troops and armored vehicles; these included the M2 Bradley Infantry Fighting Vehicle and the M3 Bradley Reconnaissance Vehicle.

We would be stationed in Aschaffenburg for two years, 1985-87. I knew about all the missions, the general defense positions for the battalion when alerted to defend from an assault by the Soviet Union. I had been exposed to all of that in the 210th Field Artillery Brigade, which was the force artillery headquarters supporting the 2nd Armored Cavalry Regiment, a forward deployed unit along the Iron Curtain between the West and Soviet Union East.

The foxhole lesson–to always do your best where you are–must not be underestimated because what you did or didn't do might show up as an essential element in a future assignment. Sometimes it seems lessons and experiences are all being orchestrated by a greater power and your only response is to do your job to the best of your ability. The rest of it will take care of itself.

Going from the community executive officer position to the executive officer of the Bradley Battalion, 1-7 Infantry, and assuming responsibilities for the personnel, training, readiness, logistics and operation of the unit, I knew what to do. I knew the general defense positions (GDP) mission. In the 210th Field

Artillery, I was around heavy equipment and vehicles and knew the general defense position locations from the role that we had there as the Force Artillery Headquarters for the 2nd Army Cavalry Regiment.

During my time with the 210th Field Artillery Brigade, the artillery officers were saying, "You're the one who keeps these things operational. You should have fired one!" And later I did! I even fired a howitzer. I pulled the lanyard on a big self-propelled, 8-inch howitzer. I was out on the range during a gunnery period, and I fired the big gun—thrilling!! This M-110 system had a firing range of over 11 miles.

Even though it took me three years to get to the infantry battalion, at the 1-7 Infantry of the 3rd ID, I already knew a lot of the things that were important. I was tracking all requirements, and the transition was seamless. Again, this was a case of having done something that I didn't think was aligned with where I wanted to go, but in the end that knowledge and acquired skill are exactly what I needed to prepare me for the next steps. And I became a branch-qualified infantry major!

By now we had been in Germany for four years: two years in Herzo Base near Nuremberg and now two years in Aschaffenburg. The whole time in Germany, Joyce and I took advantage of being in a foreign country and being in Europe. We would take long weekends and holidays with KJ and Kahni to visit neighboring countries, including travel to most western European capitals.

On a trip to Amsterdam, we made sure to visit the tulip fields. At ages 14 and 10, KJ and Kahni had traveled to European countries from Denmark, Italy, and France to England, Belgium, and Yugoslavia. They were old enough to appreciate the value of travel, and that was a huge benefit for Joyce and me, and an extra educational component for them.

Visits to unit places like Berlin before the wall came down were sobering and uniquely interesting. Other capitals left more humorous memories. As our bus traveled through the red-light district in Amsterdam, KJ and Kahni asked us why ladies were sitting in the windows. They seemed satisfied with whatever creative answer Joyce gave them at the time.

Between the move from Herzo to Aschaffenburg, my mother, my sister, a nephew, and an aunt came to visit, and we were able to expose them to Germany. They really enjoyed it, even when driving on the Autobahn throughout Bavaria. It was a wonderful four years in Germany.

We suffered through a couple of deaths as well. We lost Joyce's mother, Mrs. Mamie Hopson Lewis. Spending that great summer with her became even more precious. My maternal grandmother, Marie L. Cashen, who I spent a lot of time with during my formative years, also passed. Attending those funerals during that time was a huge comfort.

Goodbye Germany – Hello Pentagon

It was a blessed four years in Germany from 1983-87, ending a tour in Europe and the question was: *What's next?* I'm a branch qualified major now and for me the next gate would be selection for lieutenant colonel—hopefully, and perhaps ultimately battalion command.

Thankfully, I had gotten to that battalion as an infantryman, but that did not mean things would happen immediately. I learned the next follow-on assignment would be a tour at the Pentagon.

Given I had been trained as a logistician, here was another curveball coming my way, returning to an area I'd been previously exposed to. Our transfer back to the states found me assigned in the Directorate of Logistics, the Army G4. This assignment found me wondering why I was not going to the Operations Directorate, G3. I'm an infantryman—a typical place that an infantryman or a combat arms individual would go. I certainly had logistics experience through various activities as a logistician, but that was not my primary specialty.

At the Pentagon, I was assigned to the Plans and Operations

of G4; and there, I was essentially a war planner. I supported the NATO Standardization Agreement stipulations for the U.S. warfighter as the NATO Standardizations Officer. I had just returned from Germany so that was right down my alley. I understood the mission, the combat forces doing the mission and the combat service support forces enabling the mission. It was another unprogrammed marriage for me and my understanding of the total picture and it proved to be a great advantage in making things happen.

As a prerequisite to my first Pentagon assignment, however, I requested another school to better prepare me for the high-level logistics systems inherent in our national logistics activities. Up to this point, my logistics experience was at the unit and tactical level. Therefore, before taking the assignment, I needed more training. The Army agreed and I was sent back to Fort Lee, Virginia for the senior logistics course. This was a coveted course because participation was by selection only for high potential logisticians. (Remember I was an infantryman primarily – with logistics as an alternate designation).

Nevertheless, I was back at Fort Lee, going through LEDC (Logistics Executive Development Course), and it was remarkable. The course covered multifunctional military logistics knowledge. I wanted and needed this course to become the very best logistician that I could be in a new role in the Army G4.

At this point at Fort Lee, I have school-aged children—a sixth grader, Kahni, and a 10th grader, KJ. For their transition, we wanted to get them settled in a home as quickly as possible. We searched for a house quickly and wanted to get that done before I departed for TDY (temporary duty) and school at Fort Lee. We were blessed to find a community and bought our very first home in Northern Virginia. Then I was off to school while Joyce and the kids settled in. It was one of many times where the job of operating and managing our family and home fell to my lovely wife, with little assistance from me.

As usual, our military family stepped up, offering support, doing those ordinary things that we would do for ourselves and for others. They, Phil and Carol Wilkerson, helped Joyce and the kids move in – another example of the military family taking care of its own. I was off to school at Fort Lee for three months. Fortunately, the two hour drive was close enough that I could drive home on weekends and be with them.

We thought going to the Pentagon would be at least a three-year assignment. That is the way it usually works. Early in my time at the Pentagon, I was selected for promotion to lieutenant colonel—within my personal goal schedule. I was very thankful that my window for promotion was opening. I had completed the requirements, and I knew I was eligible.

The only drawback I could see was that the efficiency report from my first brigade commander in the 210th Field Artillery Brigade at Herzo Base might keep me from battalion command. Fortunately, other accomplishments I'd achieved following that report, and narratives that signaled consistently outstanding performance over a wide range of activities, got me over the wire. I was selected for promotion to lieutenant colonel while in the Pentagon and selected for battalion command in 1988, after only a year at the Pentagon.

That was a unique happening: I'm an infantryman there in the Army logistics. Typically, when promotion lists come out, the leader of any entity will send congratulatory notes to those selected for promotion within his or her element. When the promotion list came out, all the transportation officers, quartermaster officers, and the ordnance officers who were selected, received a congratulatory note from the Director of Logistics in the Army G4, a three-star general.

My boss, General Gus Pagonis, was a one-star general and director of the Plans and Operations Directorate. He knew that I had been selected for promotion and command and assumed that I had received a note from the G4 director like the other officers. However, when the director's notes were distributed,

I didn't get one. General Pagonis was confused as to why; he explained to the three-star and said, "Kip was on the promotion list to lieutenant colonel and selected for battalion command."

When the lists of selected officers are publicized, they are categorized. Typically, the three-star general has a staff member who will look at the list and select the people that are in the directorate. Being the Logistics Directorate (Army G4), the staff looked at the quartermaster list, the transportation list, ordnance lists, which included about 20 or so guys that were in the Army G4. The staff member didn't think to look at the infantry list because an infantryman being assigned to the G4 is rare. It was a simple oversight—and a teaching moment—that overlooking the infantry list, assuming there were no infantry guys here was a mistake. Logisticians could be of varying ilk's.

When the three-star Lieutenant General Jimmy D. Ross was informed of the omission, the notes had already been prepared and signed and distributed. I received a handwritten note, delivered to me personally, with him apologizing for having missed me on the first go round. I was thankful for that, and I certainly understood. But from then on until the day he left this earth, whenever I saw him, he would remember how he had forgotten. From that day forward, even after he'd retired and I'd been promoted to a general, he'd talk about that event. That note was extra special because it was handwritten from him, and it let me know the importance of a handwritten note of congratulations. I endeavored to follow that example the rest of my career.

Having been selected for lieutenant colonel and battalion command at this point meant my expected three-year assignment at the Pentagon might not happen. The battalion command list came out about 4-to-6 months later, summer of 1988, and we knew slating assignments would be out, likely for the next year.

With an anticipated three-year assignment in the DC area, Joyce's career opportunities were looming large. The kids were in school. KJ was a junior at Woodbridge Senior High and Kahni was in the 6th grade at Fred Lynn Elementary. Significantly, Joyce was slated to begin a position in Navy personnel taking her to senior executive service (SES) after GS -13/15 assignments. Performing various jobs early in her career: in civilian personnel offices, engineering and housing directorates, substitute teaching, and childcare provider validations, this position would be an exceptional opportunity to solidify her career path.

With her business degree, Joyce had been one of only two women in the business department at Morgan State. Her career was clearly something that was important for us, and I was happy for her. Although she had worked at our previous duty stations in Fort Bragg, Korea, West Point and Germany, this was a job that started as a mid-grade position, that would progress yearly as she moved forward. I was sure we'd have at least another two years in Washington, granting her opportunities to advance to senior civil service grades.

Low and behold I got activated early for battalion command. Now, instead of staying in the Pentagon in Northern Virginia for three years, I was going to leave early. Wow, it was a great opportunity; but KJ was in his junior year in high school and Kahni was beginning her 7th grade year. Joyce was in a lucrative position to begin advancing, and now I get selected to battalion command: "And oh by the way, Ward, it's Alaska and your report date is October 1988 for a November change of command!"

It's Alaska!

When the Alaska announcement was released, I was at my desk in the Pentagon. My teammates in the Army G4 teased and asked, "How are you going to go home and tell your family this?"

I was thinking, we made the decision to always go together when we did something, but moving to Alaska just might change that dynamic. I went home to give the news to Joyce and the kids. We were happy for the battalion command selection and the fact that the activation happened early—but there were mixed feelings. We had discussed going to places like Fort Campbell or back to Fort Bragg or some other continental stateside location, but now we were going to Alaska! My goodness gracious!

When the kids came home from school, we told them, anxious for their reactions. KJ—a junior in high school was ecstatic—I mean over-the-moon-happy with Alaska. Candidly, Joyce and Kahni weren't all that excited about moving to Alaska. But KJ's enthusiasm played an important role. For him, the move would seriously impact his school life, just beginning the 11th grade. He would be uprooted and wouldn't graduate with his friends and current classmates; he would join a new school in the middle of his junior year not knowing anybody. We didn't know what that would be like, but he was all for it!

KJ had one request. His big stipulation related to turning 16 in October. He asked, "Dad, can I get my license before we go?" Living in many environments, he wanted what most 16-year-olds want: independence. He wanted his license before he left, so we made sure that happened. He was exuberant.

KJ's enthusiasm caused Joyce and Kahni to cave in and say, "Well… OK… I guess we'll go." My regret was that it hindered the growth of my wife's career journey. Here was another case where the spousal employment aspect of military moves was an afterthought. So, in November of 1988, having returned from Germany in the summer of 1987, we packed up and did a permanent change of station from Northern Virginia to Fairbanks, Alaska, where I was to command a light infantry battalion. This was my first time in a light infantry outfit.

Battalion Command

Having served in an airborne unit, the 82nd Airborne Division at Fort Bragg, North Carolina, ultimately becoming a master paratrooper, also in heavy or mechanized units, including the 210th Field Artillery Brigade in Germany, the 2nd Infantry Division in Korea and the 3rd Infantry Division in Germany, I was selected for battalion command. The 5th Battalion, 9th Infantry was a light infantry battalion of the 6th Infantry Division at Fort Wainwright, Alaska. It was the first truly light infantry battalion I'd served with. As light infantry, we moved and maneuvered essentially on foot versus via airborne operations or by vehicles. However, in Alaska, we did have arctic mobility equipment, including snowshoes and skis, snow machines, and other snow mobility vehicles. The Iron Curtain still hadn't come down and the primary mission that my battalion had was protection and securing of the defense early warning (DEW) line across the northern tier, the Aleutian Islands and the associated radar sites. We rehearsed and practiced going out and protecting these critical radar installations in case of a Soviet attack. That is why we were in Alaska, in addition to being a global response force under the Army's forces command. That is why it was so important that my battalion was proficient in the use of snowshoes, skis, and other arctic mobility measures to protect and defend these remote early warning sites, if needed.

I had been to Korea in the 2nd Infantry Division where the company I commanded was a mechanized unit. My time in Germany with the 210th Field Artillery Brigade was all heavy mechanized equipment. My experience in the Bradley Battalion was another mechanized infantry outfit.

So now I go to command this light infantry battalion in Alaska. I was activated early because the commander of the battalion had been relieved. The division commander called and said, "I need a commander. I need someone here now and I need a proven special leader that can take this battalion and turn it around."

As I listened, I realized it would take work to get the battalion back to where he could count on it as one of his warfighting formations. With that in mind, we set out to report to the 6th Infantry Division in November.

Considering Alaska's cold environment, my family packed heavy clothes. We had lived in cold environments before and because it was October-November, when we arrived in Alaska, we managed fine with the cold climate. However—we learned that Alaska cold can take you to another level of cold that requires another word for describing it—a new kind of cold we had not experienced.

As I settled into taking command of this battalion, I could see many reasons why the previous commander had been relieved. The division commander explained that the battalion was a demoralized and beaten battalion. He told me, "All that I hear is that you're the guy that can turn things around and get this battalion back on track."

Once we arrived, I got my in-briefing and began to do what I needed to do to take command of this battalion, designated 5th Battalion of the 9th Infantry (Manchu).

Here was another example of how small the world is. Major Gordon "Butch" Burke happened to be in this brigade. Butch and I had met in Korea as company commanders. And we had been classmates at Command and General Staff College at Fort Leavenworth in 1982-83. He was a good friend and had a great reputation in the division.

He said, "Kip I looked at the list of potential commanders and saw your name and I told the division commander and the brigade commander that the guy they wanted was Kip Ward to come take this battalion."

I'm sure in significant measure that I got there because of what Gordon Burke said. How these things get put in your life—that is the good Lord's hand at work. Butch and I remain dear friends to this day.

Unexpected blessings surfaced, like the kids making an easy transition between schools, and even though it was not the job in personnel with the upward trajectory, Joyce landed a job that fit well with her schedule and skills.

Gordon and Rhee Burke sponsored our move and helped our transition. As military families move from post to post, this is especially important. Having sponsors makes the move so much easier. With them and their kids, we were well taken care of in Alaska. With eager confidence, I took command of the 5th

Battalion, 9th Infantry.

Brigadier General Sam Ebbesen was the assistant division commander and the senior commander in our community. The division was split in two locations: half was in Anchorage and half in Fairbanks. We were in Fairbanks. My immediate boss, Colonel Jim Wood was the 2nd Brigade commander. Our family initially lived next door to the Woods. We hit the ground running and began making immediate inroads in making the battalion the best battalion in the division. It was a great assignment.

In summer, Alaska is famous for mosquitoes and horse flies. We called them B52s because they were big and slow flying. In the summertime heat, those B-52s would invade trash piles and spread disease. General Ebbesen made a big deal about closing Dempster Dumpster doors to limit access by the flies and mosquitoes. He would say, "Make sure your Dempster Dumpster doors are closed." We knew his reference was literal and figurative.

In addition to the sanitary impact, he would use this example as a message to note how commanders and senior NCOs complied with his directives. The idea was if he couldn't trust us to do the simple things, then he might not be able to trust us to do other things as well.

In the division level training and maintenance readiness meetings, commanders would begin providing their updates; the boss would question and find holes in everything they said. Very seldom, however, did I get cross examined over some training, personnel or maintenance issue. They'd say, "Kip, how come the boss never gives you all of that grief?"

I'd say, "It's because I do what he tells me to do! This is not hard, Fellas. It is not rocket science. You do your job and what's asked of you. If the boss says close the Dempster Dumpster doors, then close the Dempster Dumpster doors!"

They'd argue, "What does a door have to do with my unit range or training status and qualifications or operational readiness rates of my equipment?" The answer was simple.

It took longer than it should have for them to recognize that not doing what you need to do when you need to do it results in all kinds of repercussions—mostly negative; and that closing the door of dumpsters was a big deal.

Later I would use that illustration as I became a senior leader commanding my brigade, division and other higher command formations. If you can't be trusted to command in the simplest of things, you may not be trustworthy to command in more complex things. When your commander tells you to do something, if it's not immoral, illegal or unethical, you do it. It may not be your priority, but if it is his/hers, then you do it. Ensuring your boss has confidence in you is a critical part of leadership.

> ***Foxhole Lesson: If you can't be trusted to command in the simplest of things, you may not be trustworthy to command or perform in more complex circumstances.***

The Push-Up Commander

I loved my soldiers and they knew it. I became the push-up guy. Personally, I did push-ups a lot and would have my good soldiers do them. Some would ask why I assigned push-ups, especially since they are considered a punishment in many circles. My soldiers knew I considered push-ups a sign of fitness; so, when I said, "Drop and let's do ten," it was because they knew I valued them as a good soldier. I was offering that person an opportunity to demonstrate physical, moral, and spiritual fitness. Doing the push-ups was our symbol of that, and that became a little trademark of mine as well. Importantly, I did push-ups with my soldiers.

On one occasion, one of my young 5 -9 Infantry Manchu soldiers returned to Alaska just out of ranger school. Ranger training is the most rigorous and stressful training a soldier can receive and endure short of actual combat. I had to get permission to send him because as an E4, normally a soldier couldn't go; the requirement stated the soldier should be an E5, but he was one of my E4 corporals. He completed ranger school at Fort Benning, came back and I was so thrilled and happy for him. I brought him in front of the formation. He was a great leader and had great potential and I had secured an exception to policy to send him to ranger school. I announced his successful completion and graduation in front of the battalion formation, and congratulated him, saying that he is a "good soldier." Because he knew my deal, he looked at me and asked, "Sir, are we going to do some push-ups?"

I didn't have to reply. When he and I got down to do push-ups, the whole battalion got down and we all did push-ups together. We bonded through unexpected episodes like that. They saw that I trusted and had faith in them. If they were out in the cold, I was out there with them—snowshoeing or skijoring behind snow machines and our small unit squad vehicles. I'm right out there with them at 4 a.m. looking up at the Northern Lights, the Aurora Borealis. They knew that I would not subject them to anything that I would not do myself. Part of bringing this battalion back from the doldrums involved spontaneous experiences like that. We trained hard. We worked hard. But I took care of them, and they knew it.

We didn't have family readiness groups per se, but the atmosphere of family support was a norm, and Joyce and I were very much into it. Our little community experienced a tragedy when one young family experienced an infant death. For many days, Joyce was right there with that family—helping them through that trauma and the grief that comes in losing a baby right after birth. Our soldiers knew that. The families knew it. The battalion was away training at the time, but Joyce had organized what was the precursor to family support groups (readiness groups now) and that tight knit entity sprang into action to comfort and assist. (While at the Army War College a few years later, Joyce was asked to participate in codifying the Family Support Group doctrine.)

As platoon leader, knowing all soldiers by name and where they were from was essential. As a company commander, I did the same thing. I figured even as a battalion commander I could know every one of my soldiers, their families and familial status –who was married, not married, those who were parents. If someone got promoted or received recognition and awards, I would write letters to their families. If a soldier acted the way they were supposed to act, I'm going to let your parents/guardians know! It might not be a parent, but I would say, "Who loves you other than me?" It might be an auntie or big brother here or there. Whoever it is, I'm going to let their loved ones know how they are doing.

By the same token, if there is the implication that the soldier started screwing up, I'm going to let these people know about it. Connections like these are what I carried on to some degree throughout my military career. I felt that sharing the news of their successes and concerns with their loved ones was important.

Training at the Army's training centers in preparation for combat was a highlight of my command tour. That rotational experience exposed strengths and weaknesses, including those made by leaders, including myself, the battalion commander. This occurred while stationed at Fort Wainwright, Alaska during our battalion rotation to Fort Chaffee, Arkansas at the Joint Readiness Training Center, where the battalion and I were put through our paces.

My soldiers knew I was out there with them. We were all sleep deprived, hungry and dirty on this very intensive 12-to-16-day rotation of continuous activity, mission after mission working towards getting proficient at our jobs.

At that rotation in the training center, going through the after-action reviews, the leader is questioned about what went wrong and what went right. If mistakes are made, the leader admits the mistake to the team and to all evaluators; there's no attempt to hide the error. That is an essential characteristic of a good leader. After all, the soldiers already know.

When I was a lieutenant colonel after one mission, I went through the after-action review and addressed the observer controller leading us, who asked, "Colonel, when did this happen? Why did the outcome occur as it did?"

I simply admitted, "Because I messed it up. I messed it up."

He said, "Yes sir, you did." That level of honesty and transparency is essential. Own the blame if it's yours.

I didn't go through a long story about what should or shouldn't have happened. I explained that I critiqued my actions and admitted that had I done this or that better, the outcome would have been different, and I would have prevented the negative outcome. And the conversation was over.

My guys smiled, clapped and laughed. Their trust and confidence were evident, and they knew that the boss knows none of us is perfect. There was no need of being unwilling to admit when I didn't make the best decision. The upside of that was witnessing how my admission created trust – they had my back, and they knew I had theirs. That trust strengthened the fabric and dynamic of our team. A big part of communication is being strong enough to own up to strengths and weaknesses.

> ***Foxhole Lesson: Admit your mistakes. Your soldiers know and will respect that. They are there and they are watching you.***

Ward School Buses

In 1989-90, I gained an experience in Alaska that emphasized the importance of soldiers having faith in me and of my showing them—individually and collectively—that I cared for them in ways they would recognize. It stressed the importance of being specific in the communication and in the actions I took as their commander.

When I was a battalion commander in Alaska, we would move by convoy or road march in vehicles back and forth from our post at Fort Wainwright to Fort Greely for training. It's about a 100-mile trip. We could take advantage of training ranges there and conduct other exercises. On one occasion I needed to go back to Fort Wainwright to participate in an event. My unit was still training in Fort Greely, so I returned to Fort Wainwright with a driver. It was late when we left, and we had about two hours driving ahead of us. It had been a long day. I was tired and I knew the driver had done what he needed to do to be ready because we had planned the trip. He knew we'd be driving back late, but I wanted to make sure he stayed awake.

As we were driving, we talked about various things. Then we kind of stopped and it got quiet. I thought, *Wow! I need to help my man stay alert and awake*. We must have seen a school bus on the road, because I started talking about this school bus company whose vehicles carried the name Ward. When my children were smaller, we'd see the buses all the time and I'd tell KJ and Kahni, "Oh, that is our company and those are our Ward school buses."

That story hung around until they were in junior high. But when my kids were small, they believed it was our company and those were our Ward school buses! He listened as I went on and on talking to him about my kids; and we made the trip fine. I didn't think very much of it. I just wanted him to know that I was paying attention—to give him that "I'm with you" comfort; so, I kept making conversation.

This episode was revived in early 2001 when I was a two star general serving as the Vice Director for Operations on the Joint Staff. Joyce and I were living at Fort Myer, near the Pentagon for the second time. A letter came. It was from my driver from that night, when I was a battalion commander in Alaska. He was a specialist (E-4) then. The letter said he was doing well after he left the Army and had started his own business in Kentucky. He remembered me and how I treated my soldiers and recalled the late night/early morning drive back to Fort Wainwright.

He said, "You started talking about these Ward school buses

and your children; I knew that was not true, but just the fact that you were taking time to talk to me about something with respect, just trying to talk to me and let me know that I wasn't alone. We were alert and in this together."

He said he'd often thought about the fact that the "old man was just talking to me to make sure I stayed awake." He said he also felt good about it because he knew I cared, and didn't take him for granted, that I saw him as a person and not just his driver.

I don't have it today but getting that special letter meant a lot to me—especially receiving it 13 years later. It was important to me to let him know that he was not alone and that we were there together. I'm glad he shared with me that it had been important to him too. Joyce never forgot that letter either. After reading it she said, "Isn't that sweet of him to recall that incident of working for a commander who cared."

I left battalion command in November of 1990, two years after taking command of the Manchus. And the battalion was now on top, one of the best in the division by all measures. However, I had to serve another seven months or so, due to my selection to attend the War College at Carlisle Barracks, Pennsylvania. I anticipated that for six months I'd be a special staff officer to the Division Chief of Staff or Assistant Division Commander.

The Division Commander, General Tom Fields, had other plans. He wanted me to be the Division G4. There was another commander coming out of one of the logistics battalions about the same time as me. In most cases, he would have been the logical selection to be the G4, a primary MOS logistician. I say logistician but I mean a real true logistician. My primary specialty was not logistics, but infantry. But the commanding general said he wanted a logistics operator with a mentality to make things happen. And I just happened to know enough about logistics to know what I was doing.

I became the Division G4 and held that position as the division went through a warfighter exercise (WFX - a division-level simulated combat exercise simulating what battalions and brigades experience at the combat training centers). I was selected "below the zone" to be promoted to colonel and shortly thereafter was selected for brigade command enroute to the War College. Guess I am a lifer!

CHAPTER 4

No Longer Due Course

From Alaska to Pennsylvania

Considered rare and competitive, a "Below the Zone" (BTZ) promotion happens when an exceptional officer is promoted to the next higher rank earlier the normal timeline gate, whose performance exceeds his or her current grade. The only BTZ in my career was from lieutenant colonel to colonel. With that early selection came my assignment to the Army War College at Carlisle Barracks, Pennsylvania, known for promoting forward-thinking concepts in military training and education. While there, I was selected for brigade command. Now, I would be a brigade commander and full colonel.

Moving from Alaska and to Pennsylvania took adjusting on many levels. I moved from commanding a combat-ready infantry battalion to an academic environment. It was like Fort Leavenworth. This position provided an opportunity and a predictable schedule that allowed me to do many things with my family.

By this time, KJ was in college at Dartmouth College in New Hampshire. In 1990, he graduated from Lathrop High School in Fairbanks, Alaska. After arriving in the middle of his junior year, he had quickly assimilated into the student population, to the point of being selected by his peers to deliver the student response at graduation.

In the fall, we put him on a plane and, unlike most college drop-offs, KJ flew solo back to the lower 48 to begin college. Joyce and Kahni sat out the trip and left me to take him to the Fairbanks airport. For them, one set of tears was enough, as we drove away from the house. Mine came as I watched him board the flight. As one friend told us, "He's moving out, and next time KJ comes home, he'll be a visitor."

Kahni was a sophomore in high school and a good basketball player. Our life and routine allowed us to do the things that dads and moms in civilian communities do with their kids. This included attending sporting events—making too much noise, shouting my child's name loudly, and giving advice to the coaches while criticizing the referee's bad calls. Kahni would look up at me like *Dad, would you stop that!* Other parents were doing the same; I wasn't alone in my antics and enjoyed being in Carlisle Barracks at the War College for that year.

The brigade I was selected to command was a part of the newly reorganized 10th Mountain Division at Fort Drum, New York. It was a storied World War II division, known for its heroic role in the Italian Alps. But to this point, given its recent re-establishment, it was not very active in ongoing missions and operations. Some classmates made comments that going to the 10th Mountain Division was not the most desirable location—nothing like the 82nd Airborne, the 101st Air Assault, or 1st Cavalry Divisions. Notably, they had not been selected for anything.

I was elated at being selected to command a combat brigade, period. And I was determined to bloom where I was planted. Yes, going to Fort Drum would be wonderful. It was unexpected, a big deal and like many promotions and assignments *we* receive, the haters surfaced. They found reasons to be critical and negative about my selection; and jealousy was there, too. It was a realization that some folks just don't like to see *us* selected for and occupying critical positions, and/or they refuse to accept that *we* are the best qualified for the assignment.

Another notable reflection of biased sentiment and attitudes surfaced, although in unconscious ways. That year at the War College, racism raised its ugly head in the form of the class graduation portrait. Each year, the graduating class selects a print to commemorate the student year; it is presented to the college by the class and prints are made available to students. In 1991-92, the picture the class at Carlisle Barracks selected included a scene featuring the Confederate flag. Some of us voiced our discontent and asked why our class chose a picture that included an image considered offensive to other classmates. I, and others, spoke up, explaining that we shouldn't select that print and that we didn't concur with it being representative of our year. But the class voted for it, never mind the undertones it symbolized.

Whether intentional or not, the overwhelming support for "A Rising Star" spoke in visible terms to the acceptance of and reverence for the cause they represented. That this print was selected for the class picture still bothers me, notably when endorsed by classmates who I considered colleagues and friends. I'd like to think that support by some classmates was unconscious, realizing they hadn't bothered to think of how that image might impact others and what it meant.

We've come a long way; however, sentiment endorsing the display of the Confederate flag is still evident in this country—and the military reflects our society. Although I've been back to the War College for conferences and to address classes, I did not specifically look for the print.

Once I graduated from the Army War College, I went directly into brigade command. I was smiling all the way up to Watertown, New York and Fort Drum to join the 2nd Brigade 10th Mountain Division (Commandos). This was upstate New York where lake-effect snow is common and expected. I had been cold in Korea, Germany, and Alaska, and now I was going to be cold again at Fort Drum. From Joyce's point of view, I had no reason to complain; she said, "Not so fast! You were never here in the cold." And in a way, that's true. No sooner than I had arrived at Fort Drum, that division embarked on a schedule making it the most deployed division in the Army.

Hurricane Andrew

During the summer of 1992, a month after I had taken command of the brigade, Hurricane Andrew struck the Bahamas, Florida and Louisiana, and our division was tasked to provide support relief operations in Florida. These operations were in support of FEMA and other federal agencies to provide relief for the devastation that had been caused by the hurricane. We were on Hurricane Andrew relief operations for 3 months—late July through October.

We provided security, cleanup, and community support for our fellow U.S. citizens. Andrew killed 44 and left a then-record $25 billion in damage. I had never seen such devastation. People were displaced from their homes, community support gone, a disaster for sure. We saw intense destruction in South Dade, Miami, and Homestead. We left Florida and redeployed to Fort Drum, thinking we were going to get back to a normal training cycle characteristic of the Army's active divisions. This would include training center rotations at the battalion and brigade levels, training exercises in the field and doing various things that units and soldiers do to maintain combat proficiency. The Commando Brigade was on task to go to the training center for a brigade war fighting training rotation at the Joint Readiness Training Center (JRTC) at Fort Polk in Louisiana. I had been in this situation before. As a battalion commander, I had conducted a training rotation at the JRTC at Fort Chaffee, Arkansas before it was re-established at Fort Polk.

As we were in preparation for the training center rotation, we began to get word of what was going on in the horn of Africa in Somalia. Soon we learned that the United States was considering the deployment of forces to support the humanitarian relief operations in Somalia. The eastern African country had been devastated by war and a deep humanitarian crisis had developed. It soon became clear that the 10th Mountain Division would be tapped to support the humanitarian assistance and disaster relief operations programs there.

On the heels of hurricane relief in Florida, we began training for the JRTC rotation, but soon training plans evolved for a mission to Somalia. On December 10, 1992, my brigade boarded C-5 cargo jets at Griffiss Air Force Base in Rome, New

York, bound for Mogadishu in Somalia. Just prior to departing Fort Drum, I was in the driveway of our home demonstrating to Joyce how to use the snow blower. Joyce says aloud—to no one in particular, "Yep! He's taking his brigade from Fort Drum to Somalia [where it averages 80°F in December], but before leaving, at 4 o'clock in the morning, we are outside, and he is showing me how to run the snow blowing machine!"

It's true! In that part of New York, snow is a big, big deal! Most of the houses on the posts where we lived came with a lawn mower for cutting grass; but at Fort Drum, residents were issued a snow blower. So, yes, we had a snow blower in the garage and before leaving for Somalia, I was out there trying to show Joyce how to start and use it. KJ wasn't there because he was in college and away. Kahni was still at home in high school. This meant leaving the two of them there to manage during heavy snows.

Joyce later observed, "We were assigned to cold Fort Drum, but in the wintertime, you were never there because you were deployed all the time—and always to a warmer climate!"

Through the full measure of this deployment, the support of the community was present and on-time. We had an Adopt-a-Platoon Program that was a part of our local Association of the United States Army (AUSA) chapter. The program existed to provide deployed U.S. military service members with quality of life and morale-lifting packages. Once the platoon left on the deployment, if families needed various things (car repair, family assistance of some kind), a community business entity would sponsor each of the deploying platoons.

With my brigade's deployment to Somalia, the local Fort Drum chapter organized community enterprises and businesses to participate in the Adopt-a-Platoon Program. That program, having started with the AUSA Chapter at Fort Drum/Watertown and the Commando Brigade, was the beginning of a program that spread across AUSA and Army.

We experienced cost-effective measures that military

families could take during deployments. For example, with two cars and one driver gone, the insurance for that car is removed, and that car is placed in storage or not driven. Families did different things to accommodate the current conditions. That was one of them, part of what we did to take better care of families.

That whole process was great when it came to furthering some of the things that we started in Alaska: family support, family readiness. Joyce was at the forefront of these activities. Continuing the work she did while we were at Fort Wainwright and codifying at Carlisle Barracks, Joyce worked with the Adopt-a-Platoon Program and guided the battalion family readiness group leaders on their projects. Now that I was in command of a brigade, we made sure that home station support structures were in place so that the deployed soldiers could focus on the mission knowing that their families were being well cared for.

Operation Restore Hope—Somalia

We deployed to Somalia for this mission labeled Operation Restore Hope, a deployment that early on met with relatively high success. Although it was classified as a humanitarian mission, neutralizing armed gangs riding on "technicals" (typically pick-up-type trucks with machine guns mounted in the truck bed) led by warlords, engaging in numerous firefights, skirmishes and ambushes made it truly a combat mission. The attack helicopters (AH-4 Cobras) that deployed with the division were often called in for direct fire support. The indirect fire weapons, mortars and machine guns were routinely used in our operations.

Given my pattern of constant circulation in my Brigade Area of Operations, the Brigade Command Sergeant Major (Command Sergeant Major Reubens) and Operations Officer (Major Dave Wood) made a quick decision to assign an M-60 machine gunner to me for personal security, saying my pistol and M-16 rifle that I also carried were not enough. On more than one occasion I was glad that a soldier (Sergeant Reynolds) with his machine gun was with me. The enemy was everywhere and knowing who they were was virtually impossible until the shooting started.

This type of combat engagement was prevalent in certain sectors of our area of operations. After the deployment, the Army and DOD acknowledged that soldiers having engaged in these combat operations would be eligible to receive the Combat Infantry Badge (CIB), given all the regulatory requirements for awarding the badge were met. As we understood, the initial reluctance was due to the mission having been classified as humanitarian.

My soldiers and I were in combat with an enemy equipped to kill us as their mission. Nevertheless, although acknowledging the mission as a combat operation that warranted the CIB, that authorization was never extended for service given in this hostile theater. No bronze stars were authorized, but peacetime awards were presented instead for this tour of duty: Army commendation medals and meritorious service medals. At some point, should the Army reauthorize those medals as bronze star medals, that would be the appropriate recognition.

Nevertheless, our overall success rate was high. The brigade was able to neutralize the impact of the armed gangs and secure humanitarian routes, allowing NGOs and other relief organizations to deliver supplies to starving Somalis.

The belligerents were the warlords and their clans and sub-clans who were fighting for control over territory. Using their technical, they were stopping the effective delivery of relief supplies by hijacking convoys and terrorizing the traditional village clan elders (leaders.) Our mission was to neutralize these groups and support the traditional clan leaders in governing their regions.

The warlords didn't want us there and one of their tactics was to set explosive ambushes along convoy routes. We didn't have the 'up armored' vehicles, and we would put sandbags in the floor of our vehicles to try to harden them. We improvised to do things to make our mission safer. These types of actions were occurring from December of 1992 to May 1993, when my brigade headquarters rotated out. My brigade headquarters, 2d Brigade 10th Mountain Division and two of my battalions returned to Fort Drum from Somalia in May 1993. I still had another one of my battalions there so I would go back to Somalia from time to time through the end of September 1993, for a command visit with that battalion.

Suffice it to say, it was a combat environment with combat operations. Mortars, gunfights, roadblocks, ambushes all were part of the operational theater. All of this was done in the name of keeping lines of communication and supply routes open so that needed humanitarian supplies could be delivered to a starving Somali population, victims of a 3-plus year war between competing clans.

These were populations that were in dire need of support and assistance from the United Nations. We were there to provide the secure environment that would allow this humanitarian assistance to be delivered. But then there were the warlords taking advantage of the chaos that existed. When the UN Force seemingly chose a side, it evolved into a series of actions eventually leading to the loss of the U.S. Black Hawk helicopter crew.

Another lesson revealed in that deployment: the military couldn't do all that is needed to bring about and maintain a stable society. The military could establish security, but it could not provide other pathways to stability, like economic and social development and the governance aspects needed. After that assignment, my first realizations were that the U.S. military could do its part, but other entities with committed and long-term involvement would have to do their part.

Volunteer organizations provide support, nongovernmental actors and organizations are crucial, but economic and social entities of that country must be part of the solution. Taking visible steps to improve people's health, food security, education and basic living standard is an essential element in creating and maintaining a secure and stable environment. (I later characterized this phenomenon: that people must have a *horizon of hope*.)

One visit to my units was in one of the small villages and my NCO was in charge of the unit, along with his squad. My squad leader says, "Hey sir! We have brought peace here. We're keeping these warlords from fighting. We've separated them, but these village elders need some help to re-establish their ability to grow food."

There it was! A small but essential infrastructure requirement that we didn't consider and couldn't support. There was no Army entity for that. For the stability and space that we created to be sufficient and permanent, other needed to come from governance and developmental components. We understood governance and development like any person, but it was not our job to provide it. Nevertheless, it was essential for them and their well-being, in whatever way it made sense for them.

That was my first realization after the fighting: that preparing communities for peace meant securing a combination of more sustaining aspects of life. Safety and security needed to be established, and we had stopped the fighting. But now a variety of humanitarian needs arose that required attention: economic advancement; social advancement in health and education; effective governance (based on their tradition and customs) so that the people knew who their leaders were and what those leaders were doing on their behalf.

> ***Foxhole Lesson: There is no physical security in a region whose people struggle without food, social, health or economic security.***

That would be reinforced in successive assignments. Somalia was my first exposure to how to respond to a community we had been protecting, as this young sergeant said, "Other things needed to be happening here."

Somalia gave me firsthand witness to understanding this. In the Port of Marka, Somalia, I would watch Somalis trying to load cows on ships to float them out to sea to put them on a freighter because one of the things they were doing was exporting cows to Europe for beef. Other things, such as rules and control at the makeshift port, effective financial systems, and accepted operating guidelines were needed to make it work. Somalia opened my eyes, and I certainly saw that later as well.

Black Hawk Down

My Commando Brigade and two of my assigned battalions redeployed from Somalia in May 1993. My headquarters was relieved by 10th Mountain Division's other infantry brigade headquarters, and my third battalion was assigned with them for that deployment. During this timeframe, the overall mission control came under the auspices of the United Nations. The belligerents perceived favoritism by the UN, with the notion that the UN had taken sides. The clan-led group that had become out-of-favor retaliated by attacking coalition forces. [The Habar Gidir was the primary clan involved in the Battle of Mogadishu, led by Somali warlord Mohamed Farrah Aidid.]

By that time special operating forces had also been introduced to the theater with classified missions. While conducting one of these specialized missions, the helicopters were attacked by the out-of-favor clan group. My battalion, now working for the other brigade headquarters, was decisively involved in the rescue operation for those downed aircrews. One of my soldiers in the 2d Battalion, 14th Infantry, Sergeant Reid, lost his arm during that mission. I was able to visit him during his recovery at Fort Meade in Maryland and present him with the Purple Heart later that year.

Just coming up on my first year in command and readjusting from the deployment, the Department of the Army (DA) alerted me that I would be asked to sit on boards. These DA boards selected officers for promotion, schools and command assignments. I was happy to receive this request because it was a central part of ensuring the best-qualified officers from across the force were given an equal chance to compete for these assignments. It reminded me of the War College class portrait situation featuring the Confederate flag and reinforced the importance of having a voice and being a part of these processes.

In early 1994, my second year of command, I was asked to serve on a board to select battalion commanders. Because being on this board meant leaving my brigade for 30 days, my division commander didn't want me to sit on that board. Learning this, I called another senior officer that I had great respect for and asked, "Sir, I've been asked to go back and sit on this board, but my division commander doesn't want me to do it." I explained that my commander felt the time away from my duties would not be good, that my duties to my unit needed to be put first.

I was happily relieved when the senior officer encouraged me, saying that it was important that I was present on those boards. "Your presence is representative of the Army," he said, "because when it comes to selecting folks to come behind you, your perspective and understanding is important and needs to be there."

So I returned to my division commander with this announcement: "Sir, if I can't leave this brigade for 30 days to go sit on a critical board, when I have three battalion commanders, a brigade executive officer who are all lieutenant colonels; and if we can't trust them to carry on the work of this brigade in my absence, then you probably need to fire me. Integral to commanding this brigade is my role in training them to be commanders, to come behind me."

They were all outstanding officers. He just kind of looked at me with a long sigh. The bottom line—I sat on that board. The brigade did just fine while I was gone, and I was able to perform a very important role in selecting the next generation of officers for a critical assignment.

I recognized that my presence on these boards and in these forums was critical. Yet someone was trying to keep me from doing that. It was my boss! Someone that I worked for and who knew me. My commanding general didn't want me to participate. That stung. Thankfully he didn't prevent me from serving on that board, but that told me something about him as well, and the type of things done that impact these processes. This was another occasion in my career where my boss treated me in a way that was plainly prejudiced. Recalling my difficulty in getting the assignment following CGSC and the first two block OER in the 210th Field Artillery, this episode reinforced my realization that my treatment was not always on a fair and equitable basis.

Warfighter Training

Warfighter training is the learning process extended to soldiers. Warfighter exercises train the command and staff teams in six areas: command and control, movement and maneuver, intelligence, targeting processes, sustainment and protection.

Although I was nearing the end of my brigade command tour, the brigade still had not been through a combat training center, Battle Training Center (BTC) or simulated warfighter rotation. The training had been planned prior to the deployment to Somalia. The Army's goal was to expose the Commandos to a BTC/warfighter experience. With this as the rationale, my brigade was selected to be the Light Brigade Headquarters for a heavy (armored or mechanized) division warfighter training at Fort Bliss, Texas.

Warfighter exercises train the division and brigade command and staff teams in a variety of combat missions; it's a substantial war game exercise that divisions go through. For the battalions and brigades, it can be an actual field training exercise or simulated exercise. Units can be deployed to the field and train through tactical exercise; and unit headquarters' staffs can train and be tested through simulations. The Commandos would deploy its staff and conduct our training in a simulated environment. Soldiers go through the entire processes and procedures of war planning, orders preparation, and execution of missions.

A Black Book Assignment

While in Texas with my brigade staff conducting this warfighter as a light brigade headquarters attached to the division, I received a call from the Pentagon. My name had gone into a *black book.* These are unique assignments for officers who are worthy of being considered for a critical, special assignment. I was being considered for the executive officer to the Vice Chief of Staff, Army. *Whew! My name was in the black book as a candidate to be the Executive Officer to the Vice Chief of Staff of the Army, General John Tilelli.* I also found out later that a War College classmate of mine involved with this process had attempted to sabotage my being included in the black book for consideration for the position. (Another reflection of what goes on in the careers of some.)

Up to this point, about to relinquish command, I had no idea what I would be doing next. And given what had been the case with my division commander, I had no support from him as to what would be my next assignment. Most former brigade commanders (FBC), having performed successfully, were assigned to specific jobs where the experience of being a FBC is desired.

With that information, I was set to leave Texas immediately for an interview at the Pentagon. Having deployed in my combat gear for a tactical training exercise, I did not have appropriate clothing. I needed my dress uniform. I called Joyce back at Fort Drum: "Hey Baby! Could you pack my uniform and send it to Washington, DC. I'll be flying there for an interview to be the XO to the Vice Chief of Staff of the Army."

It was now early June 1994, and I was scheduled to change command in July, but didn't have a job. Calling Joyce with this news, at least we knew that I was on the radar for a significant position, although no guarantees. Spouses and families serve too, and this impacted her. That something was brewing mattered, and you'll never tell me that I did not see and hear Joyce's smile over the phone! She did her part and more, as usual, by getting the uniform together and sending it to Washington. She performed admirably, doing what needs to be done to help and support the family. Indeed, the families are the Army's combat multipliers.

The interview at the Pentagon with General Tilelli went well from my perspective; however, I wasn't sure I'd been selected for the job until a few days later. The Division Chief of Staff called me and said, "We have to move up your change of command, General Tilelli wants you to be his XO when he becomes the Vice."

My successor in the Commandos had been selected: Colonel Jim Dubik. Jim and I had been lieutenants together in the 3/325 ABN Battalion of the 82nd ABN Division, and I knew him well. We coordinated, planned and conducted the change of command in a record four days.

As it happened, all the speeches were short due to a thunderstorm that came up. The soldiers were happy for a short ceremony and given that I had to be at the Pentagon the following day, we all ran off the field as soon as Jim and I passed the brigade colors. Remarks were short: Jim said he was happy to be there, and I thanked my soldiers, families, teammates from across the division and citizens from the North Country/Watertown, New York community who were great supporters and had become good friends. My only regret was not having more time to more formally thank all who had made this command tour so rewarding. After shaking hands, I went home, grabbed a few things and drove down to Washington, DC to assume my duties as the Executive Officer to the Vice Chief of Staff of the Army.

Earlier that spring, Joyce and I were able to take our daughter, Kahni, who was in her senior year of high school, on some college tours as she finalized her decision. She was a high school varsity basketball athlete, and although she had received several scholarships to play basketball at other schools, she decided she wanted to attend James Madison University in Harrisonburg, Virginia that coming fall.

Thankfully, KJ was also with us having just graduated from Dartmouth College in New Hampshire in June 1994. Joyce and the kids packed up our house at Fort Drum a few days after the change of command to follow me down to Fort Myer, Virginia, near Arlington National Cemetery and the Pentagon. The moving van showed up a day later on Fort Myer and we—basically Joyce—moved on post to our designated housing. That was not the only time she took on the lion's share of moving logistics, and it would not be the last. Another testimony to the resilience of military spouses.

The position as XO to the Vice was an education about the Army and how the Army runs that can only be learned from that position, and General Tilelli was an excellent teacher. Because of my boss, I was involved in everything the Army did: from personnel matters, operations, and logistics to finances and budget for the Big Army. I witnessed it at the highest level, operating on the E Ring of the Pentagon—another foxhole.

Joint Assignment

Having been selected for general officer in 1994 while serving as the XO of the Vice Chief of Staff, I knew my first assignment as a brigadier general would be to a joint billet. During those days, newly selected flag officers not already joint qualified received a joint assignment as their first assignment as a flag. The Secretary of Defense had determined that all general and flag officers (GO/FO) should know more than their basic service.

Being assigned to a joint position gave officers knowledge and understanding of the other services (for me, Air Force, Navy, Marine) in our defense department. Thus, if an officer had not been assigned to a joint billet where they were able to know other services, their first assignment as a GO/FO would be a joint job. The value is that the position allows someone to then serve at higher levels with joint responsibility, possessing the requisite experience to handle the position. In the fall of 1994, my name was posted on the brigadier general's list. That meant I was now a promotable colonel, and I knew that I would only be on the Army staff for a year.

In 1995, after one year of being the Executive Officer to the Vice Chief of Staff to the Army, General Tilelli promoted me to brigadier general: assigned to the joint staff, as one of five Deputy Directors of Operations (DDO) in the J3. This assignment provided a great education into the joint world and the global employment of the joint force.

I had worked with joint forces in Somalia as a part of the Joint Task Force for Operation Restore Hope. Being on J3, the staff gave me an opportunity to really understand joint force in ways that I hadn't understood before. As a DDO, I functioned as one of the nation's five general officers on rotational duty, in charge of the National Military Command Center (NMCC). We were the nation's watch officers. Each of us commanded a team of service members from all services who were in constant communication with the U.S. Defense Forces wherever they were deployed around the world. We were the direct liaison between the nation's warfighting forces (combatant commands) and the Secretary of Defense and Chairman of the Joint Chiefs of Staff.

Typically, five teams rotate in the NMCC. Each team is headed by a one-star general/admiral from the Navy, Marine Corps, Army or Air Force. I was the team leader for Operations Team 3. We worked rotating shifts. The continually rotating schedule could be tough: sometimes working days, sometimes working nights, sometimes working the mid-shift.

I spent 16 months in that command center and it was my first predictable experience. A person would work four days on, two days off, in 8-hour rotating shifts, never working the same shift back-to-back. In working the night shift, coming off the night shift, typically midnight to 8 a.m. the next day; would lead to a four-day break. Three teams worked this 24-hour shift with the other two readying for duty or just coming off duty rotation.

I loved it! It was the first time in my military career that I had a predictable schedule. I knew when I was working and I knew what the schedule was because when I was not in that seat, another general or flag officer just like me was in that seat. When one person left, it did not mean the work ended. This was a 24 hours a day, 7 days a week operation and all included knowing what is going on all around the world. The National Military Command Center was tied to every combatant command, every service, to all the national intelligence operations—comprehensive and inclusive knowledge of everything.

A significant outcome of this predictable schedule came when, after many attempts, Joyce and I could finally plan and follow through on something special: we could plan our 25th wedding anniversary. A couple of times, we had planned to take a cruise; something came up for me and we had to cancel. We tried for another cruise after returning from Germany, but Joyce had started a new position. With the issues spouses experienced in pursuing careers and not wanting Joyce to miss that opportunity, we canceled that cruise.

Now this is spring 1996 and our 25th anniversary is approaching. It was a predictable time for me; we planned a cruise. I went down to the travel office in the Pentagon and organized yet another cruise. This one stuck! Joyce and I took our first cruise for our silver wedding anniversary while I was there at the Pentagon. Our 25th anniversary cruise was a wonderful, wonderful time. Who knew an Army man would enjoy being on the water so much! When people talk about cruises, I'm the first to say, "Buddy, you're going to love it!"

As time progressed, I gained great appreciation for the joint services, all services, and what they bring to our national defense. I was familiar with Naval Carrier Strike Groups, destroyers, amphibious ships and big decks, air wings, fighter wings and where they were. I knew the Marine expeditionary forces and divisions and gained great appreciation for our Special Operations Forces. Of course, I knew the importance of Army Formations, Brigades, Corps, Divisions and Sustainment units and what they bring to bear in securing and dominating territory. I developed a heightened sense and appreciation of all these complimentary elements that would serve me well in my subsequent joint assignments. I truly recognized and honored the value that is brought to bear by the joint team.

In late summer of 1996, I was looking forward to leaving the Pentagon following my first brigadier general job. I was a soldier, having been promoted to general officer and although I needed that time with the joint team, I wanted to go back to an Army division. As a one-star in an Army division, the job that we all want to have is that of assistant division commander (ADC); today they are called deputy commanding generals. We all wanted to be an ADC in any one of our Army divisions.

The soldier in this ADC position is not the person in charge, not the commander. The ADC interacts with the troops and has oversight over various elements of the division. The division typically has two ADCs: one has oversight for maneuver operations, the other has oversight over all the enabling, sustainment and support activities that make up the division. The ADC's presence matters in providing direct oversight as an extension of the division commander.

As I left the joint staff to go to my division assignment, I had another professional development program to attend. There is an education program for *"baby generals"* called Capstone, an orientation class for newly appointed generals and admirals. This is part of the military's institution education process. NCOs have institutional schooling from basic to primary leadership development schools to Sergeant Major academies. Officers have them as well: basic courses, advanced courses, intermediate course command and general staff college, the senior service course or War College.

Capstone was a four-week program, part of professional military education. Newly promoted one-star generals and admirals learn specifics about the Navy, Army, Air Force, Marines, and spend time in DC with the inter-agencies. This was an extensive course that provided learning space for these general officers to know about our national systems. Airborne and Special Operations capabilities are substantial components, therefore orientation to these activities was part of Capstone. One of the destinations for the Capstone group was to Fort Bragg.

Korea Again—Or Not!

While attending my Capstone course, hoping to go to a division to be an ADC, I was notified that my assignment would be to the 2nd Infantry Division to be the assistant division commander in Korea. I had served in Korea as a young lieutenant and captain and had commanded a mechanized infantry rifle company in Korea. I knew what that mission was, and I knew about Korea. Situationally, it was an Army division forward deployed and that was going to be fine for me. In preparation for our move to Korea, we arranged for the packing and shipment of our household goods.

Sometimes Army circumstances change right up to the last minute. It can easily be the case that while packing up and nearly in route to one assignment, another assignment comes that supersedes the first and plans change. Our household goods were loaded for shipment but hadn't been shipped when we got the notification: "You're not going to Korea. You're going to Fort Bragg."

In preparation for the assignment, I had exchanged notes with the 2nd Division Commander in Korea, but that was now overcome by events. My assignment to the 2nd Infantry Division as ADC in Korea was changed to the 82nd Airborne Division, at Fort Bragg, NC. The reasons for this change would soon be revealed.

In 1996 when I returned to Fayetteville, the KKK billboard had been removed. Still, this was 1996, and racial/extremist problems existed in the military and at Fort Bragg. The Installation Commander was dealing with issues of right-wing extremist groups; body markings and tattoos reflecting various gang and extremist affiliations were surfacing in all branches of the services. Gang-related activities that drew unspoken and divisive lines within units were celebrated. At Fort Bragg, there had been a shooting by one of these extremist individuals as soldiers were assembled for morning physical training in one of the athletic stadiums on post.

That there were no minority general/flag officers on this large installation, Fort Bragg, was cited as a reason adding to the polarization. There may have been one or two lieutenant colonels or colonels, but no minority general officers. Fort Bragg included a large population of soldiers in the 82nd Airborne Division, other units of the 18th Airborne Corps and Army/Joint Special Forces.

The senior leadership of the Army in the Pentagon and on Fort Bragg were fully aware of the environment on Fort Bragg, and this environment was degrading the combat readiness of the assigned units. Subsequently, a decision was made to diversify senior leadership. They understood that a difference in leadership was needed as the environment was negatively impacting readiness. I was tapped to be the ADC for Support in the 82nd Airborne Division and help deal with the turmoil that was going on.

Obviously, I was well qualified and prepared for this assignment. I had no special instructions along with the assignment; however, as an ADC, I was visible throughout the division, the largest unit on Fort Bragg. If nothing else, my presence was visible and it mattered. I was the only Black general officer at Fort Bragg. And this was not just with units of the 82nd Airborne. I was visible every morning throughout the division area doing physical training with every unit on the famous Ardennes Street. Already a master paratrooper, I conducted parachute jumps weekly, jumping more often as a general than I did as a lieutenant, 25 years earlier. My presence was known and felt across the entire installation.

Issues of race and diversity were with us then and they are still with us. These are things I had been exposed to, and while progress has been made, issues still remain.

Being the ADC for Support in the 82nd Airborne Division at Fort Bragg was a great tour for me. In the division, there are typically two of us one-stars. As ADCs, we rated the division's brigade commanders and separate battalion commanders. We had oversight across the division. It was a good experience for me. I was mentoring juniors. Working with battalion commanders and brigade commanders - some of whom went on to achieve great things in our Army - was rewarding. A couple of them had been my students when I was teaching at West Point.

My time at Fort Bragg also came with some sorrow. As I was coming off the parade field after having been the Commander of Troops at the division's change of command, my 'aide de camp' came and told me that Colonel George White had passed away in Pennsylvania after going there for specialized medical treatment. We called him "Champ." He was a good friend, I loved him dearly. Our families were close. We had met in 1979 when I was stationed at West Point as an assistant professor in the Department of Social Sciences.

I went over to the new division commander, Major General Keith Kellogg, who I had just conducted a change of command for, and said, "I have to leave. I just lost a friend and I'm going to be the family's survivor's assistance officer." Joyce and I drove to Hampton, Virginia to provide some comfort, support and strength to the family of a great friend.

The rest of the tour at Fort Bragg went well. From ensuring the high level of readiness and sustainment of the 82nd Airborne Division to performing myriad activities for the Department of the Army, I enjoyed my time. Separately, General Hugh Shelton was commanding the relatively young Special Operations Command in Tampa, Florida. He had previously commanded the 18th Airborne Corps at Fort Bragg and was aware of my presence there. He reached out and requested I make a TDY trip to Tampa and address his command. Naturally I agreed and by all accounts had a productive and valuable leader professional development (LPD) session with his joint command. I appreciated the relationship I had with General Shelton.

Two-Star Major General

In 1997, while at Fort Bragg in the 82nd Airborne Division in my second brigadier general job as the ADC, I was selected for major general (two-star). Although I was enjoying this tour in the 82nd, with the promotion I knew I would be leaving Fort Bragg. The question was, "to where?" As a career infantryman, commanding one of the Army's divisions would be wonderful, and while a few new combat arms two-stars may go directly into division command, that was not to be my case. Instead, pinning on in early 1998, I left Fort Bragg for an assignment as the Chief of the Office of Military Cooperation in Cairo Egypt.

I had no prior exposure to that type of assignment and wondered what it was all about. I realized: it was another one of those jobs that you do because you are told to do it; you do it to the very best of your ability. Joyce and I packed our household goods, left Fort Bragg, North Carolina and shipped out to Cairo, Egypt.

I learned the mission to Egypt was to lead the U.S. defense cooperation activities with Egypt emanating out of the peace treaty in 1978 between Israel and Egypt. The main actors in that historic circumstance were U.S. President Jimmy Carter, Egyptian President Anwar Sadat, and Israeli Prime Minister Menachem Begin; they received the Nobel Prize for Peace for this in 1978. The treaty also included a request to the United Nations to establish the Multi-national Force & Observers

(MFO), an international peacekeeping contingent to maintain peace between the two former warring parties, primarily in the Sinai Desert.

The U.S. provided defense support to Egypt through an appropriation of 3 billion dollars a year. That is the program that I oversaw, but it was separate and distinct from the MFO mission. Although the MFO mission was not my responsibility, as the senior U.S. Defense Department official in Egypt, I felt it a responsibility to routinely check on the U.S. component of the MFO mission. At the time, a young Captain Johnny Davis was an infantry company commander there. He and I quickly established a relationship and maintained it throughout the years.

My office was in the U.S. Embassy in Cairo. Although I was not an official member of the Country Team (the organization in our embassies under the authority of the U.S. ambassador with representatives from the Dept. of State and other federal departments and agencies), I regularly coordinated with the ambassador and attended his Country Team meetings. Working with U.S. Ambassador Daniel Kurtzer, I built relationships and gained an understanding of the activities of U.S. departments and agencies in the field. Dan was a tremendously talented and effective diplomat. He and his wife, Sheila, became dear friends to Joyce and me.

Because of my experience in previous assignments, by now my default was that when I got a job that I knew little about, I would get educated and do research. That is standard for most people going into a new environment—learning the inner-languages, observing cultural preferences and social expectations. I became more familiar with the inter-agencies in the embassy in Cairo and became familiar with other groups I'd be working alongside. In addition to the State Department, I became familiar with the U.S. Agency for International Development (USAID), Treasury, Commerce, Agriculture and other departments and agencies.

Because of the unknown and what things might lead to, the

smart thing was to perform the assigned job at the highest level, and the rest would take care of itself. That is what happened with my time in Egypt working with the Egyptian military and the Egyptian intelligence services. This was a very complex environment, politically, militarily and socially, particularly regarding the Sinai peacekeeping mission and Egypt having to deal with the Muslim Brotherhood creating internal conflict. The country includes a sizable Coptic Christian population who were sometimes threatened; and there were criticisms of a lack of democracy.

Foxhole Lesson: Go into a new situation with respect. Be open-minded and eager to learn before making major decisions or judgments.

Being the senior U.S. military official in Egypt in 1998 was a wonderful assignment. Joyce and I were excited to be living in a part of the world we had only read and heard about. We were on the continent of Africa—seeing and visiting the pyramids, the Sphinx, traveling to Upper Egypt.

While we were there, we were able to enjoy visits from both of our children. Joyce and I lived in the Egyptian economy, and I learned a bit of Arabic, mainly from Osman, my driver. Our house in Cairo was in a section called Maadi. The house was designed for entertaining, and it was useful as I had representational responsibilities as the senior U.S. military official in Cairo.

In Cairo we visited many historical sites, including the Great Pyramids of Giza, the Sphinx, the Valley of the Kings and Queens, ancient tombs, and Luxor along the Upper Egypt Nile. Some visits were at the urging of the then-president of Egypt and supported by our U.S. ambassador. Special attention was paid to the security situation. We rode camels and Arabian horses around the pyramids. Climbing the Steppe Pyramid, known as the Pyramid of Djoser, which stands 205 feet tall, was a uniquely gratifying experience.

I worked closely with the Egyptian military and Minister of Defense. My responsibilities allowed me the opportunity to know President Hosni Mubarak. My comfort in these high-security relationships came from an understanding of and respect for different cultures. For Joyce and me, taking a crash course in Egyptian customs, culture and traditions prior to this PCS paid off in innumerable ways. The training helped me see the value of getting out of my foxhole to understand things from someone else's perspective. That helped me see the world from the point of view of those I was working with! You'll understand people and situations better, and you'll be able to act more effectively. Realizing that was a big part of that process, and I really enjoyed it.

Working with the ambassador as a part of a country team where the other parts of our government are represented was quite an experience. From working with USAID, the Commerce and Treasury departments, and various represented agencies, all were part of the *country team.* It was a big learning experience for me and exposed me to a lot.

In this assignment, I developed a firm grasp of how the USAID, if its development mission is done effectively, has a huge role in enhancing stability in a country. It was a pile on to my Somalia experience, notably that sustained peace is obtained through populations being cared for in ways that matter to them, through the mission of more than Department of Defense activities. But—as rewarding as the Chief of the OMC mission was—it still wasn't commanding a division.

> ***Foxhole Lesson: Learn the value of getting out of the foxhole. Look for opportunities to see the world from someone else's point of view.***

From Egypt to Hawai'i

The Egypt assignment was my first two-star job and about 19 months into that job there in Egypt I was notified that I had been selected to command a division. My goodness gracious! The Army has 10 divisions and I'd be commanding one of those divisions. Wow. There are only 9 other individuals doing what I was about to do. As it turned out, I was about to command the division I had visited for a research project when I was a student at the Army War College (Carlisle Barracks, Pennsylvania).

In 1992, as a student at Carlisle, I had gone to Hawai'i to interview the then-serving division commander, Major General Fred Gorden. Now, seven years later, I'm headed back there to command that same division. So once again, Joyce and I found ourselves packing up, this time headed for Hawai'i to command the Tropic Lightning Division, the 25th Infantry Division there in paradise.

The division had a storied Vietnam legacy and a succession of great commanders, from Fred Weyand to Harry Brooks to Fred Gorden, each of whom I would get to know. Some criticized the Army and the U.S. military in Hawai'i because of some of the things that had been done historically and how it occurred. This criticism was based on damage done to the land used for our training areas, etc. Mākua Valley is a local training area that Army and Marines use to conduct live fire exercises. It was a part of what I was responsible for as both the Commanding General of the 25th Infantry Division, but also the Commander, U.S. Army Hawai'i.

When I arrived in Hawai'i at Schofield Barracks, home of the 25th Infantry Division, one of the Vietnam-era division commanders and former Army Chief of Staff, General Frederick C. Weyand was still living. By then he was retired, but I got a chance to interact with him and we became friends. He would call me his *commander!* In fact, when General Weyand received the Marshall Award from the AUSA (The Association of United States Army, the Army's premier lobbying organization), as he received his award he said, "I just want everybody to know my commander is also here, the best commander that this division has ever had, Kip Ward," and he added that included him!

Joyce and I sat there flabbergasted! His comments were totally unexpected and out of the blue. His acknowledgment told me something about advocacy. He didn't have to do that, but felt it important to acknowledge me in a way that was very public with leaders from all over the Army. It reinforced for me what I had learned about advocacy reflecting on what Chuck Hood did for me. Yes, even at that level, advocacy matters.

> ***Foxhole Lesson: When you are in position to advocate for someone who might not otherwise have an advocate or a voice, and if that person deserves it, you need to say something. You need to do that.***

Respecting Hawaiian Culture

While commanding the 25th Infantry Division, I learned about the principle of *Mālama*—a Hawai'ian word with a deep meaning—*to care for, protect, preserve.* The word includes having a spirit of respect and responsibility for community, the environment and for self. A Hawai'ian environmental group, *Mālama Mākua* , was present and sensitive to what the U.S. military was doing in the Mākua Valley training area. I made it my business to go out and meet with them to try to understand their concerns so that our operations were taking those concerns into account. It was our way of observing and listening, to see things from the perspective of others. I wanted to hear from my Hawai'ian friends and neighbors.

Their concerns were basically about our training and how our training activities were detrimental to their sacred lands. The question was how would we accommodate that. Mālama Mākua would meet in the early morning around 4 a.m. or sunrise and participate in ceremonies. With their permission, I joined them on occasion. A Mālama ceremony honors the land and encourages participants to share love, respect, and compassion. Understanding that, my response was: *We are going to be respectful of your land just as the ancient Hawai'ian warriors were when they trained here.*

I explained that we would have to train here, but we would not do it indiscriminately and we would do it in a way that is respectful of the land. That made a great impact. In fact, when I left, before Senator Daniel Inouye passed, he would say, "Kip, they are still talking about you and how you paid attention to them and ensured the land was protected. They knew you were sincere in your concern."

My concern and respect for their culture were genuine; that was so important for our relationship. Taking the time to listen, understand, and get to know people, and then being respectful of what is important to them goes a long way in establishing good communications and strong relationships. In doing so, we got along. It wasn't perfect, but the Hawaiian caretakers of that valley knew that we were trying to do the best we could. They understood that as the commander, I was meaningfully engaging with them and not just giving them lip service in trying to work things out. That was a meaningful part of my Hawai'i experience.

My brigade had been part of a division warfighter training exercise when I got tapped to go back to DC to be interviewed for the Executive Officer to the Vice Chief of Staff of the Army position. Now I was a division commander about to participate in my own warfighter in the 25th Infantry Division, this very important training exercise.

Other than being in actual combat, the division warfighter training is considered one of the best tests for success in a division. The warfighter exercise was a simulated training exercise, and participating in this training was as close to real as possible. It tested the division commander and staff in combat, combat support and combat service support missions of the division in offensive and defensive operations.

So important was the division warfighter training exercise that the Chief of Staff of the Army would be present for the final after-action review of the warfighter. On that day, he listens to critiques by the evaluators. The commander and staff respond to scenarios and input from the retired four-star general, who is there as a senior mentor to observe/evaluate the command posture and actions of the division commander. The senior mentor, a retired four-star general, is selected by the Army Chief of Staff as the overall evaluator. In this case, General Rick Shinseki selected retired General Edwin Burba. General Shinseki came out from DC as the Chief of Staff of the Army to attend my division's final after-action review following the warfighter exercise.

My overall senior evaluator, General Burba, a former commanding general of the United States Army Forces Command (FORSCOM) in Atlanta, was known for his sternness and toughness. I did my job as the division commander, just as I had done in my battalion at the JRTC at Fort Chaffee, Arkansas and in my brigade for warfighter training in Texas.

My Division Chief of Staff and all but one of my brigade commanders went on to be promoted to brigadier general. I had a great command team and division staff. As General Burba pointed out to General Shinseki, "Ward provided the type of leadership that let that team be as he had trained them, and excel."

That warfighter evaluation process reminded me of what folks had done for me and how I processed and used that. From advocacy to letting my team lead, I gave them direction while trusting them to perform without micromanaging. (Later I learned that General Burba told the Chief of Staff how effective the division's operations were. He was complimentary in acknowledging my leadership of the division, my making the correct decisions based on staff input, and my intuition in making the right things happen.)

The division command tour was a good command tour. It wound up being shorter than I wanted. We successfully accomplished the warfighter exercise in seldom realized efficiency by all standards, and I was ready to keep humming along, training for our mission in the Pacific theater and taking care of my soldiers, families and civilian workforce. Somehow, with the success the division achieved in warfighter training and the overall positive trajectory of the division, General Shinseki made a comment as he was leaving to return to the Pentagon that gave me a feeling that my tour might end short of the two-year norm. Sure enough, within a couple of weeks I received notification that I was going back to the Pentagon.

General Shinseki knew about having a division command cut short. He was a division commander when I was executive officer to the Vice Chief of Staff of the Army. I watched him come out of his division command early to return to the Pentagon to be the Deputy Chief of Staff, Operations in the Army G3 as a two-star. I was sorry he came out of the division early because division command is such a big thing. When I saw him in 1994, I expressed how sorry I was that it had happened. He said, "Well Kip, you go where the Army needs you and you go and serve and do the best that you can."

And now it was happening to me. I was coming out of division command early to go back to the Pentagon to be the Vice Director for Operations on the Joint Staff (the Vice J3). Y2K had come and gone without incident (after much concern).

Joyce and I packed up for my assignment back to the Pentagon. Ironically, General Hugh Shelton was now the Chairman of the Joint Chiefs of Staff. I later discovered there had been a discussion about me and how the Army needed to nominate the right officer for this critical billet.

Army housing would not be available on Fort Myer for several months, so Joyce and I moved into temporary living quarters on Bolling Air Force Base (in D.C.), waiting for our designated quarters to be ready. Within a few days I was settled in my office at the joint staff. My position included participating in interagency staff meetings at the White House, and I could be called to go there and over to Congress any time of the day for security meetings. I kept uniforms in my office; that was just part of the job, a job that included coordinating our national defense response to crises, deploying the global Army, Naval, Marine and Air Force units and other interagency coordination. So, I always had extra uniforms in the office.

On September 11, 2001, I had just returned to my joint staff office from the NMCC at the Pentagon when the plane hit. My office was on the opposite side of the Pentagon from where the plane crashed. Even so, we were all impacted and knew something horrific had occurred as alarms were sounding throughout. I didn't leave the Pentagon for five days afterwards.

Being in a building when an airliner flies into it is incredibly sobering! Knowing that it had happened, unable to leave, being in there for five days, and continuing work was intense. It took a while before Joyce heard from me and knew that I was okay. Fortunately, about 3 or 4 hours into the crisis one of my staff assistants was able to call her and say that I was fine, just really—really busy.

By now we were living on Fort Myer, adjacent to Arlington National Cemetery, so Joyce could see the smoke coming from the Pentagon. Our daughter, Kahni, worked in DC and could look across the river and see the smoke. Our son, KJ, was in Boston at the time. Friends from all over the world and family, who knew that I was in the Pentagon, called Joyce asking about me. For a while she had to say she didn't know until she finally heard from one of my staff.

While all of this was going on I was on a mission—coordinating with the Federal Aviation Administration and the North American Aerospace Defense Command to clear the airspace over the United States; to deploy forces to protect critical infrastructure and facilities in the U.S. and around the world. This was the beginning of the global war on terror and

Operation Noble Eagle, the name given to all North American Aerospace Defense Command (NORAD), aerospace warning, control, and defense missions in North America. It still goes on today and puts the air cap over selected airspace. Through Operation Noble Eagle, NORAD deters, detects, and defeats potential threats to U.S. and Canadian airspace 24/7/365.

Experiences from 1995 and 1996 when I was Deputy Director for Operations in the NMCC came into play and proved invaluable. Now as the Vice J3, the NMCC worked for me. I knew it well because I lived it. So, when it came to coordinating across our agencies, NORAD, the FAA and controlling airspace, scrambling aircraft to go up to intercept and patrol airspace or whatever needed to be done—I was up to the task. I worked along with the National Security Council to determine what our policy directives would be for interdicting a threatening aircraft (even a U.S. airliner), if in fact it came to it.

Dad's WWII Engineer Unit, 1942

Joyce's parents, Levi and Mamie Lewis, and my parents, Richard and Phyllis Ward

My grandfather's church Falls Road AME
Church, Baltimore MD, 1949

Me with school-age children, Falls Road Village of Cross
Keys, 1957

Defensive Back #30, Towson Sr High, 1967

With Sergeant First Class (SFC) Mattison and Ward, 1969

Appointment as Bear Brigade Commander with Morgan Interim President Fraser and PMS COL Hurt

Morgan Reserve Officer Training Corp Cadet Ball, 1971

Soldiers Armory, (me at top right), centered is Brigadier General (BG) Chappie James, 1971

Field 1LT Pin-on by Lieutenant Colonel Lyle Alexander, 1972

325th Airborne Infantry Battalion Officers, Fort Bragg, NC 1972

Korean Demilitarized Zone (DMZ) Patrol, CPT Ward 1975

Company Commander, C Co 1-17 IN 1975

Me and 1SG Hill in Camp Howze Korea, 1975-76

"We're going to war!"

Instructions to Platoon Leader Jim Stratton, 1975

Airborne (ABN) School 1971 and 82nd ABN
Farewell Jump (C47) 1998

KJ
in
KOREA
1975-1976

Me and KJ, Camp Howze HQs, 1975

Me and KJ, visit to Yongsan Korea, 1976

Joyce and KJ, Apartment in Yongsan, 1976

Ward Family, Graduate School - Penn State University 1977

KJ and Kahni at West Point, NY, 1980

CHANGE OF COMMAND
5TH BATTALION, 9TH REGIMENT

2ND BRIGADE
6TH INFANTRY DIVISION (LIGHT)

NOVEMBER 2, 1988
FT WAINWRIGHT, ALASKA

MANCHU HISTORY

Part of the pride of an organization arises from its history and traditions. The 9th United States Infantry Regiment, known by associates and comrades in arms as the "Fighting Ninth" and more recently by the additional title "MANCHU", is a proud organization rich in organizational history and in traditions as illustrious as those of any combat unit in the history of warfare. Indeed, the history of the MANCHU Regiment is entwined with the saga of every major conflict of our existence as a sovereign nation. Its traditions arise from the first organization in 1789, the War of 1812, the Mexican War, the Civil War, the Indian Wars, the Spanish-American War, three distinct tours in the Philippine Islands, the Boxer Rebellion, World War I, World War II, and more recently the Korean and Vietnam Conflicts. The MANCHU Regiment as a unit, and as individual men, distinguished itself in each of its 62 campaigns. As a result, the Ninth Regiment earned the respect and admiration of the nations with which it has served, and its many opponents, inspiring awe with its numerous deeds of valor, acts of courage and moments of self-sacrifice.

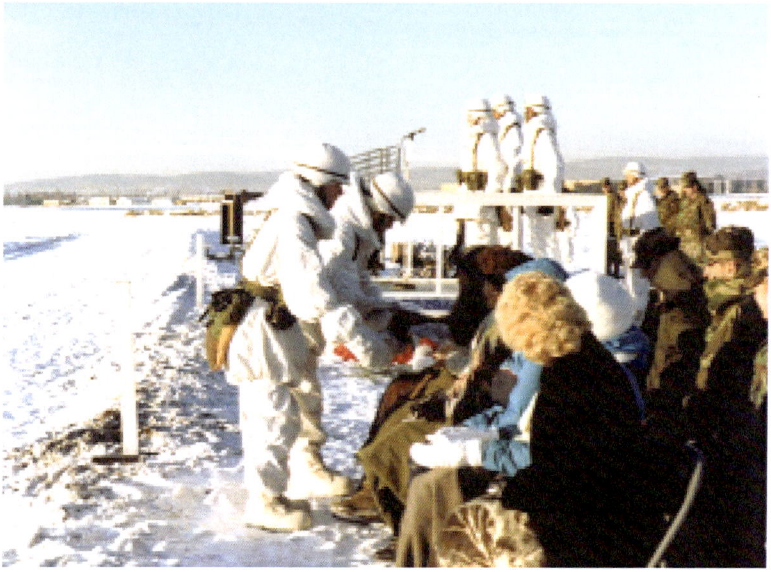

5/9 Infantry Battalion (Manchu),
Change of Command Ceremony at Ft. Wainwright, AK, 1988

Manchu Battalion Command Run, Fairbanks AK, 1989

Joyce with women of the Manchu Family Support Group,
Ft. Wainwright, AK, 1989

Manchu Battalion on Parade, Ft. Wainwright, AK, 1990

With Jim Wood and Randy Bookout conducting traditional Liscum Bowl Ceremony by Manchu commanders in 1989 at Fort Wainwright, Alaska. The Manchu slogan is, "Keep up the Fire!"

My dear friend George "Champ" White, 1989.

Best friends George "Champ" and Rita White, 1991. I left the parade field in 1996 to be the Survival Assistance Officer for Rita and their boys.

With Sgt Reid, Ft Meade Hospital 1993

With Chairman, Joint Chiefs of Staff Powell, Somalia, 1993

With Norty Schwartz on 1996 Capstone trip to the Pacific

Capstone stop in Japan, 1996.

Commander of Troops, 82nd Airborne Parade,
Fort Bragg, NC, 1996

Joyce Ward and her husband, Brig. Gen. William E. Ward, talk about their military experiences at his office at division headquarters.

Staff photo by Marcus Castro

Rising star puts soldiers first

By Henry Cuningham
Military editor

"The bottom line is I am still a soldier.
Knowing my soldiers has always been

Joyce reflecting on 26 years of marriage to a soldier who loves his troops, 1997.

With retired Legendary "ROCKS" Generals in 2000--Smith, Becton, Leigh, Honor, Price, Klugh, Gaskill

Preparing for Farewell Jump leaving the 82nd Airborne Division in 1998 with senior leadership. Leornard Price conducting the Jumpmaster Preinspection safety check.

1997-8 Fort Bragg memories.
Top left: Family Christmas portrait at home.
Top Right: Dinner with Geoff and Pamil Miller.
Bottom Left: Dennis Via presents farewell gift on behalf of 82nd ABN Div Separate Battalion Commanders.
Bottom Right: With mentee Bruce Moody

Joyce's farewell by Paige Kellogg from 82nd ABN Division Family Readiness Group, February 1998.

Cairo American High School graduation ceremony, with Joyce and KJ, Cairo, Egypt, 1998

With Egyptian Naval officials and my U.S. Navy program chief, Alexandria. Egypt, 1999.

Top: With Osman, my OMC driver
Bottom: Joyce with Feli, staff assistant at
our residence in Cairo, 1998

With OMC-Cairo personnel.
Top: Gary Gariepy, my C-12 maintainer and good friend
Bottom: Joyce with Betty Gariepy and Family Readiness
Group spouses at our residence, 1998

With Ambassador Daniel Kurtzer and Secretary of Defense William Cohen, 1999, Cairo, Egypt

Joyce with Janet Cohen and Suzanne Mubarak, 1999. Cairo, Egypt

Joyce unveils new plaque naming Naval Medical Research (NAMRU-3) in Cairo Egypt, 1999, with Hafez Labib, Edward Antosel, Sherif Azmy, and Rich Haberberger

OMC-Cairo Air Force (F-16) and Army (Apache Gunships) training assistance teams, 1999.

Christmas in Hawai'i with the Wards, 1999

Kahni and KJ, Schofield Barracks, Hawai'i, 1999

Major General Ward, 25th ID Commanding General (CG) Hawai'i 1999

Wards in Hawai'i– Quarters 227 Generals's Loop, 2000

Former 25th Infantry Division Commander and Army Chief of Staff Fred Weyand saying goodbye to Joyce with Dave Hodne and Jager looking on in background and farewell to Tropic Lightning Division Troops as we leave Schoefield Barracks, Hawai'i, 2000

Command Group after 25th ID 5-mile Run (L to R, CG MG Ward, DCG BG Tucker, C/S COL Kelley, and DCSM Ripka)

Celebrating the Triple Nickles in 2000

With Dick Benjamin (top) and Joyce and Kahni with Liz Benjamin (bottom) on our 2000 visit to Germany from Egypt

With friends Greg and Lisa Packer at US Army Europe (USAREUR) Formal, 2004

Governor Doug Wilder, 2016 (together in Jerusalem in 2005)

In the Oval Office with President George W. Bush, 2005

With Joyce and Secretary of State Condoleezza Rice at the United Nations, 2005

With Israeli Prime Minister Ariel Sharon, 2005

Farewell Ceremony, leaving USAREUR and heading to EUCOM, 2006

Our May 2006 4-star promotion ceremony family portrait with mom, sisters Tina and Dorothea (Douglas), brother Levi (Paula), our children KJ (Randy) and Kahni (Kenyatta), aunts Hilda and Doris, nieces, nephews and cousins.

At my 4-star promotion ceremony with our unwavering advocate and supporter, Arthur Gregg. A true legend. 2006

With Deputy Secretary of Defense Gordon England (to my left), EUCOM Staff on either side of us, and Supreme Allied Commander Europe Gen Jim Jones (far left) at Gen. Jim Jones's Farewell, Stuttgart Germany 2006

AFRICOM Deputies AMB Mary Yates and VADM Bob Moeller in Burkina Faso, 2007

Hon Colin Powell remarking on the significance of me becoming the nation's first Black combatant commander in standing up AFRICOM in 2007

Klaus and Gabrielle Kloessel-Schaefer, Stuttgart
Friends, 2008

Renewing my realtionship with Bono in 2014

AMB Mark White, early Senior Development Advisor at AFRICOM

With Triple Nickle Troopers (555th Parachute Infantry Battalion),
Air and Space Museum, Washington, D.C., 2015

Collage, including Ted Koppel and President Clinton in Bosnia-Herzegovina

Top: With President John Kufour of Ghana
Bottom: With the Ghanaian Chief of Defense Staff in Accra, 2007

CSM Ripka speaking at Uganda People's Defense Force NCO Academy, 2009

With friends Vimla Lalvani and Conrad Schloesser in Germany, 2008

US Africa Command Staff and Senior Leaders Offsite, 2008

With US Ambassador to Liberia, Linda Thomas-Greenfield arriving in Liberia

With Rwandan President Paul Kagame, 2008

t to right, Chairman of the Joint Chiefs of Staff Admiral Mike Mullen; Henrietta H. Fore, administrator of the U.S. Agency for Internatio
velopement and director of U.S. Foreign Assistance; Defense Secretary Robert M. Gates; flag bearer; U.S. Africa Command Commanc
neral William E. Ward; and AFRICOM Command Sergeant Major Mark S. Ripka stand together after the unfurling of the flag during t
RICOM Unified Command Activation ceremony in the Pentagon Conference Center, Washington, D.C., October 1, 2008. (Department
fense photo by Petty Officer 2nd Class Molly A. Burgess)

AFRICOM Activation ceremony as a unified command, 2008

Ward with *Black Enterprise* Publisher Earl Graves in Stuttgart, Germany, 2010

With Berlin Airlift key figures, Ms. Mercedes Wild and Col Gail Halversen, and Juanita Harvin, 2009

I always took time to engage with local community leaders and military officials in thoughtful dialogue to foster mutual trust, understanding and collaboration.

Visiting my Navy Seabees from CJTF-HOA in Comoros, 2009, 3Ds at work

Speaking during Pentagon Press Conference, 2009

EUCOM and AFRICOM staff members Tom Cloma,
Gerald Hunter, Brandon Alexander. 2009

AFRICOM Portrait

With Rwandan Chief of Defense, Gen James Kabarebe in Kigali, 2010

Representing the Army at the Essence Music Festival, 2010

Asst. Secretary of State Johnnie Carson, 2010

Testifying before the Senate of Armed Services Committee, 2010

On Mission at AFRICOM

Retirement ceremony and review with 3d Infantry, The Old Guard, Fort Myer, VA, 2011.

Retirement Ceremony Memories at Fort Myer, VA 2011
Top Left: Major Luckett
Top Right: Mark and Deb Ripka
Bottom Left: Garry Bishop
Bottom Right: Steve Parker

With GEN Austin at my 2011 Retirement ceremony

With longtime mentor, Johnnie Wilson. Morgan State University Presidents David Wilson and Earl Richardson in background.

Joyce and me with President Barack Obama and First Lady Michelle Obama at the White House, 2011

Ambassador Thomas-Greenfield concludes her tenure as US Ambassador to the UN, 2025

Joyce and me with Ambassador Thomas-Greenfield at her residence in New York, 2024

With Steve Harvey at his Boys Camp, 2011

Recognizing AFRICOM civilians and service members for their exemplary work, Stuttgart, Germany, 2011

With trailblazer Julius Becton, 2007

CGSC Class 1982-1983, reunion cruise 2012

With Mo Ibrahim at the Africare Memorial Dinner and Gala, 2013

Joyce and me with Earl Graves, 2013

Joyce and me with Peter and Soo Pham at the African Arts Awards, 2016

Joyce on the Morgan State Bridge in 2013, remembering where we started dating.

Wards at the Trumpet Award Alumni Program, Atlanta GA 2014

West Point friend (79-82) and CGSC Classmate (82-83), George Ingersoll in 2016

Joyce and me with Frances Hesselbein in NY, 2016
(met Frances in 2004 in Germany)

George White's (Champ) Boys Andrew, Rashad, and Brandon, 2016

With Master Warrant Officers (Ret) Fred Norman and
Eddie Royal, 2017

Fellow 1993 10th MTN Div brigade commander Fred and Esther Valenzuela, 2017

Me and Joyce with Butch and Rhee Burke at his retirement from Govt, 2017

Joyce and me with Susan Kawazoe and her cousin in 2018. Susan was my secretary in Hawaiʻi commanding 25th Infantry Division (Tropic Lightning) in 1999

Christmas Gala, 2018

In Des Moines for World Food Prize panel with former Nigerian President Obasanjo, 2018

With Former USAID Adminstrator Mark Green in Maputo, 2019

Emma Sky, 2005 POL-MIL and Development Advisor in Israel, 2018

Joyce as a champion for childhood education and family readiness with Dr. Jill Biden, 2015

With President George Bush in Dallas TX, remembering Israel and the stand-up of US Africa Command, 2023

At Morgan State University with Tuskegee Airman, Charles McGee, 2019

Present for remarks by Arthur Gregg at the historic renaming of Fort Lee to Fort Gregg-Adams in 2023. Reversed in July 2025.

Fort Leavenworth CGSC 1983 classmates at our annual reunion, 2023

DEAR GENERAL + MRS WARD
WE WANT TO WISH YOU + YOUR FAMILY A JOYOUS HOLIDAY SEASON + A HAPPY
Wishing you the happiest of holidays
AND HEALTHY NEW YEAR!
MAY GOD BLESS YOU + YOURS + KEEP YOU WELL!
WITH FONDEST REGARDS + ALL GOOD WISHES!
DORIS + JACK GOLDMAN

Holiday card from Doris and Jack Goldman

Joyce with my former AFRICOM staff executive assistant Susan Patrick with her husband Ed, and my former admin officer LTC Lyrica George taken 2015 in Hawaii

My childhood home. It took 6 years, but my father built this house with his bare hands, brick-by-brick.

Triple Nickles reunion with Sam Boyd and
Joe Murchison

Joyce and me celebrating wedding anniversaries with Pat and Erv Pearson

Stuttgart, Germany collage

CHAPTER 5

Three Stars and Back Across the Atlantic

After 9/11, the United States looked to avenge that attack by taking action against those who had perpetrated it. I stayed busy in the Pentagon six or seven days a week: working 18-hour days on weekdays, and 12-hour days on weekends.

On Sunday I had a little routine. Joyce and I lived at Fort Myer, near the Pentagon. I'd get up early on Sunday morning, go into the office, get some things done, and go back home, in time for us to attend the 11 o'clock chapel services. After lunch, I would go back to the Pentagon for a few more hours and leave around 5 p.m. Sunday evenings were my time off. When the weather was good, Joyce and I would go someplace to just sit and talk; otherwise, we'd be there at the house catching up. That was our routine throughout that year.

The United States had planned and began prosecuting the Global War on Terror (GWOT). My dad passed away in 1999 while we were stationed in Cairo, Egypt. Thankfully, my mother was still with us. Living in Virginia, we were fairly close to her home up in Baltimore. She often noted that I saw her more when I was overseas than during this tour at the Pentagon. Then I was a major general, commanding the 25th Infantry Division in Hawai'i and now, as the Vice Director of Operations on the Joint Staff, it was so hard to get away—virtually impossible given the ongoing global operations involving the U.S. Department of Defense Forces.

While in my third assignment as a two-star general, I was nominated for my third star. (It is a bit unusual to have multiple assignments in the same general officer grade as was my case in every grade from brigadier general to full general.) The nomination and subsequent promotion to lieutenant general came with a new position. Once an officer moves above the rank of major general—the last permanent rank—subsequent promotions are based on specific duty assignments. An officer must have a job that authorizes them to perform at that grade.

For my rank, I was nominated and then confirmed to be the next commander of the Stabilization Force (SFOR) in Bosnia-Herzegovina-Herzegovina (BiH). This was a position organized and operating under the North Atlantic Treaty Organization (NATO). That assignment came with a promotion to lieutenant general (three stars). The Secretary of Defense and President of the United States made my selection and nomination for lieutenant general. The confirmation was made by the Senate. This is the process for all senior generals and admirals, three- and four-star level. I would begin that mission in November 2002.

It was a yearlong assignment and categorized as a " hazardous duty" assignment, meaning another unaccompanied tour for me. Joyce would not be going to Bosnia-Herzegovina. We considered whether she would stay there living on post on Fort Myer, Virginia.

In addition to the U.S. force structure assigned to this NATO mission, the U.S. Army provided much of the logistical and enabling support for the mission. When the mission first started in Bosnia-Herzegovina, the commander of U.S. Army-Europe was also the commander there. Going forward and for subsequent commanders, it was established that a set of quarters in Heidelberg, Germany where U.S. Army-Europe was headquartered would be designated for the commander in Bosnia-Herzegovina as well.

Stabilization Force in Bosnia-Herzegovina

Although I would not be living there, Joyce and I were authorized to do a permanent change of station and move to the house in Heidelberg. Because I was in and out of Bosnia-Herzegovina for visits to Brussels and other European capitals, we saw opportunities to be together whenever I would come to Heidelberg. She had lived in Germany before and enjoyed it.

I was already in Bosnia-Herzegovina as Joyce prepared for the move, determining what would be stored and what would be shipped to Germany. Having to manage that change alone was another challenge facing military spouses from time to time. Because the service member has already departed or already engaged in the next assignment, spouses are often alone to manage the actual move.

When Joyce arrived in Germany, I was able to meet her, however I couldn't stay very long. I spent two to three days as we worked to get her settled in. Initially she was at a temporary place until the actual house was ready. Once our household goods arrived in Heidelberg, she moved in and set up the house while I was in Bosnia-Herzegovina.

It was another three or four weeks before I could get back to Germany again to be a part of the process of setting up the household, but that was okay. We were in the same time zone and could talk frequently. I knew that she was being well cared for and things worked out well. Military families take care of each other through formal as well as informal programs - many of which Joyce spearheaded over the years. We had friends in Heidelberg. We knew one couple, Bill and Juanita Harvin, from our first tour in Germany in the 80s.

Joyce always says, "We have our blood families, but our military families are also very special and always there." So often military families are with you in places where your blood relations are not. We have that kind of relationship with many, many families.

In Heidelberg, Joyce was right at home and found additional connections with many people in that community. Some relationships resumed and renewed, and we made some new friends in the process. The Pearsons (Erv and Pat) and Masons (Jim and Yolanda) are dear friends of ours till this day and we celebrate anniversaries and weddings whenever possible.

Around that time, we met the Packers—Greg and Lisa and their sons, Gregory, Jr. and Matthew. Joyce and I became godparents to their boys, and they are our military family for life. Both boys are graduates of William and Mary. Today, Gregory is a lawyer, and Matt is a member of the U.S. Secret Service.

Operation Joint Forge in Bosnia-Herzegovina

Even though I was not assigned to USAREUR (The Army Service Component Command for Europe), U.S. Army-Europe, the U.S. Army headquarters for all U.S. Army forces in Europe; they provided wellbeing and quality-of-life support for people living in the military community. There were many people with U.S. Army-Europe that I knew and/or remembered from previous years of service.

With the Stabilization Force, Operation Joint Forge in Bosnia-Herzegovina, I was privileged to be the operational commander for forces from 24 NATO nations and other troop contributing nations. My headquarters was in Sarajevo, and I had troop elements across the entire country of Bosnia-Herzegovina. The 24 NATO nations and other non-NATO members, including Russia and Morocco, were aligned with the mission.

The experience of being in a joint environment in the Pentagon gave me an advantage and I leaned on my understanding of different services and cultures within the Army, Navy, Air Force, Marines, and Coast Guard. Aside from being in Germany, in Hawai'i I had worked with (but not commanded) Indo-Pacific partners. Here in Bosnia-Herzegovina, I was the commander of troops from Italy, France, United Kingdom, the Netherlands, and other countries.

It was important for me to understand the cultures and various preferences of the local population. I needed not only to understand the teammates in my command, but also to operate effectively in Bosnia-Herzegovina, where the war had occurred and three different cultural groups lived among one another: the Croats, the Bosnia-Herzegovinaks, and the Serbs; and each represented different religions: Muslims, Orthodox Christians, Catholics, and Jews. All that dynamic and more was part of Bosnia-Herzegovina.

In addition to that, our command provided support to the European Union High Commissioner, a civilian British official named Paddy Ashdown. That meant another important dynamic had to be considered: managing civil and military relations.

As the leading military official in Bosnia-Herzegovina, my role had equivalency to the civilian role performed by the High Commissioner of the European Union, who was responsible for the civic activities going on there. Understanding those elements and working in that complex system of myriad national interests, protocols and cultural variations was critical. Add to that the unique dealings with United Nations participants (including the Secretary General) and the European Union (multiple member nations and their governments) activities, from policing to building back a society that had been devastated by war. Indeed, this was a complex and tense situation, compounded with a dynamic set of circumstances.

Merging Military Components in a Stable Society

When it comes to stability, I was given great exposure to how the military component works hand in hand with other elements of a society. It's a complex, many-sided cube including the economic, political/governance, development and the social aspects of health, education, and religion. These elements must work together when it comes to building and maintaining a stable society. Clearly the security piece was vitally important; however, the substantial issue was in understanding that whatever security we established then would not guarantee the long-term stability of a society if those other elements were not being addressed.

In numerous ways, being the commander of the NATO's stabilization force in Bosnia-Herzegovina-Herzegovina exposed me to the importance that these elements have in collaborating and partnering to truly make a difference in creating a stable society. How we related to one another, how we cooperated and coordinated and synchronized our actions—those were critical for moving forward.

Essential to my mission was acknowledging that I didn't need to understand their jobs to the degree that I would be able to do their jobs, but I needed to understand that what I did was in harmony with what they were doing. I became sharply attuned to that, understanding the limits of what I did and how my security mission helped to create a more stable society.

That experience was beneficial later in Israel with the Palestinians and Israelis, in the United States European Command (EUCOM) as the Deputy Commander, and in Africa as I worked as the inaugural commander of the United States Africa Command (AFRICOM). I considered that assignment a major part of my development and continued learning as a soldier in valuing different perspectives.

Foxhole Lesson: Take advantage of everything that is presented so that you learn and understand that well. You never know when that knowledge or skill might serve you well later. Seeing things from outside of your foxhole makes for better decisions.

As the commander of a multi-national joint combined operation with security operations going on, my mission included searching for the persons indicted for war crimes and apprehending war criminals who were still at large. It was a hazardous duty assignment. The war criminals were still operative. We were operating in an environment that was unfriendly while we worked to re-establish security and stability. We were a part of the U.S. Defense Commission that was established to define the defense structures in Bosnia-Herzegovina. We did this in coordination with various officials. All of that presented complexities beyond basic infantry operations. This was truly an environment where tactical actions had strategic consequences.

I made trips back to the United Nations, rendering reports to the UN Security Council, to Europe to brief the European Union partner, to Washington for consultations, and naturally to Brussels for briefings at NATO headquarters with my boss, General James Jones. With a Russian airborne brigade assigned to SFOR with responsibilities in a critical sector, a trip to Moscow also occurred.

In Bosnia-Herzegovina, nations were assigned sectors based on the alignment of these nations with the population. The NATO mission attempted to align forces with sectors having populations familiar to them, culturally and otherwise. So, there was a Russian brigade working in the Serb sector of BiH conducting force protection operations.

Clearly nations take actions in their own interests, so the Russian interests were different from ours, as were those of the Turks and the Italians, French and British. But if nations work together to find those common lines, that helps bring connection, stability, and peace and that is what we all wanted.

Commanding SFOR was truly a career defining experience. The command was comprised of rotating forces with most coming to Bosnia-Herzegovina for a six-month period, including my U.S. contingent. Some of them were there for a bit longer. My command tour of duty in Bosnia-Herzegovina was for one year, and a tour filled with many critically important experiences.

One initiative was bringing together the heads of various religious groups. I had met one or two of them at a function and as we talked, they provided me with very rich information on culture and history. I wanted to meet with them on a periodic basis so I could continue to learn from them. They thought that was impressive. From their perspective, they saw a U.S. general in charge of this NATO force who wanted to learn from them!

They were really taken with that and that initial meeting evolved into a luncheon with the religious leaders of the major religions, every month at my headquarters in my command mess. The group included: the Mufti from the Bosnia-Herzegovinak Muslim community, the Croat Catholic Priest, the Serbian Eastern Orthodox Christian leader, and a lay leader from the small Jewish community in Bosnia-Herzegovina.

They looked forward to our monthly meetings, where their suggestions and ideas surfaced and where they could freely collaborate with each other. When they saw that I was truly interested in learning from them, they came together and spent that time. To this day, folks think that was such a unique arrangement, that I had these meetings with religious leaders whose beliefs and doctrines often clashed. I could take what I learned from each of them and help guide my actions so that as I interacted in the society and did what I did in my command it made sense there, as opposed to something that would alienate them or be viewed as not aligned with things that were important to them.

I noticed similar dynamics at work with the high representative from the European Union, Paddy Ashdown. He and I spent time together as I ensured that my activities were aligned with his objectives as high representative. I can't emphasize enough how important it was to synchronize those pieces, and it had a big impact when it came to intergovernmental relations even within our own government when we set up AFRICOM. Understanding your partners and what they are doing adds value to the total effort and does no harm to the work being done by others. All this is essential for building a more stable society.

> ***Foxhole Lesson: Get out of your foxhole, go downrange and see your position from someone else's viewpoint - it'll make for better decisions with likely more buy-in from your partners. Understanding***

> ***your partners and what they are doing adds value to your effort and reduces the potential for undermining the objective.***

My work was enhanced by and with these various components of SFOR. There was an Italian Carabinieri unit that I got to know well. The Carabinieri are Italian paramilitary forces, organized as a military unit and charged with maintaining public security and order as well as assisting local police. They are a versatile force. There was an Italian brigade, and I was invited to Rome to their Carabinieri headquarters as their special guest. I wanted to make sure that I was employing them in a way that made them most effective and in accordance with their operating modalities.

As my most versatile force, that proved to be a good move and showed my specialized Italian unit that I wanted to understand how I could best employ them as opposed to just directing them. They appreciated that and as it turned out I became a bit of a novelty personality there amongst the Italian Carabinieri.

> ***Foxhole lesson: Good leadership means understanding that those under your leadership are not interchangeable. Developing an appreciation for the modes, priorities, and strengths of the people working with you allows you to engage them in ways that are both most aligned and rewarding for them and most effective in the overall cause.***

The Push-up General

On one occasion in Bosnia-Herzegovina, the Carabinieri unit conducted an airborne operation and one of the members got blown a little off course. Being a paratrooper, I understood that. After recovering from having landed in the wrong place, the Carabinieri soldier finally made his way to join up with us. I went over and talked to him. He was fine and we laughed and joked about the situation, then I said, "Alright! Let's do some push-ups buddy."

Now I was known as the "push-up general," carrying on my tradition as a battalion commander. My reputation was that I did push-ups with my good soldiers. I was proud of him because after getting blown off course, he collected himself, got things together, and joined the mission ready to perform.

It is important to recognize that things won't always go the way you want, but when you can regroup, get it together and carry on, that is something to recognize and celebrate. So, in recognition of him dusting himself off and getting back into position, we celebrated by doing push-ups and we did that together. The press happened to be around and thought that was such a novel thing. Everyone jumped down and started doing push-ups with us!

My point was that I do push-ups with my good soldiers. Not only is it a sign of a soldier's physical fitness, it's also a sign of their moral fitness, spiritual fitness, and commitment and dedication to the mission.

Foxhole Lesson: Do the things with your soldiers that resonates with them and conveys your intent.

As I was leaving Bosnia-Herzegovina at my change of command event, High Representative of the EU Paddy Ashdown, commented, "Ward goes around, and he does push-ups with his soldiers who do good work." He turned to me and said, "I thought I was a good soldier, but you never asked me to do pushups with you!"

So there at my formal change of command ceremony, we did push-ups together. Although the religious leaders never did push-ups with me, they'd always remind me, "Hey, we're your soldiers, too!"

Impressed by their sentiment, I'd laugh and say, "No, no! You're not my soldiers, but we are on the same team." That was important because it showed how you work together as varying groups of people, but all towards the same mission.

Horizon of Hope

I enjoyed the tour and the people of Bosnia-Herzegovina. I came to understand they want the same things for their children as we do: to live in safety, to live in peace, have relatively good health and good schools. It was really a sign that people are no different all over the world.

By and large we all want essentially the same thing. Clearly persons and entities arise that would do self-serving and nefarious acts, but when you talk to and about common everyday citizens, they want peace, security, stability, reasonable health and a decent life for themselves and their children where they are. They are rarely looking to go elsewhere. They want it where they are. So, our mission was to try to provide an environment where they could achieve that right where they were, and I tried to carry that everywhere I subsequently went. It's what I began to refer to as *having a Horizon of Hope – the idea that the prospect of something on the horizon can be achieved that will enrich my life and the lives of those important to me in ways that matter.* Taking the time to talk and listen to people who are different was invaluable in completing that mission.

Conversely, there were those who benefited from the chaos, who did things to undermine efforts for peace and stability. This happens in many places and it's why working together is so important. Sometimes when things happen that disrupt their ability to create chaos, they don't like it. Bosnia-Herzegovina was my first time having a security detail. I went on to have one assigned to me from just about then on for the rest of my career. Because of the activities I was conducting in Bosnia-Herzegovina, I was disrupting a set of actions that were being taken by bad actors. Those bad actors did not want their activities to be disrupted because they were achieving some personal gain from their misdeeds. This created an inherent security risk to me personally.

My security team was with me everywhere I went. I got to know the women and men who were a part of this protective detail extremely well because of the time we spent together. My personal security details included all American soldiers. This was my close security. If I were to go to another area or sector, there would also be security provided by SFOR forces who were assigned in that area.

As the conclusion to my one-year tour was approaching, the question became what would be my next job. One possibility was to be the commander of the 18th Airborne Corps at Fort Bragg. I was interviewed by the Secretary of Defense for that job. I had recent service in two of the corps' three divisions: the 82nd Airborne Division and the 10th Mountain Division. Following the interview, I was told I was going to be selected for that job.

Instead, an intervention occurred with another officer selected to command the 18th ABN Corps. While I had looked forward to returning to Fort Bragg, other factors changed that assignment, as had happened to me before. I was selected to be the Deputy Commanding General of the U.S. Army Europe. The effect of that meant keeping Joyce and me in Europe and ironically, as she had moved to Heidelberg, it was the headquarters of U.S. Army Europe. And again, with trust in God and doing my best in whatever job given, things ultimately worked to my favor.

Deputy Commanding General, U.S. Army Europe

So, I moved from Sarajevo as the commander of the SFOR in Bosnia-Herzegovina to Heidelberg to be the Deputy Commanding General (DCG) of U.S. Army Europe. Even though we stayed in Germany and Heidelberg and on the same installation, we had to do another PCS move.

The house that Joyce lived in while I was in Bosnia-Herzegovina was a dedicated home for that commander in

Bosnia-Herzegovina. But the house that the DCG lived in was a different set of military quarters. So, we moved from one house down the block, around the corner and up the street to another house. We had another PCS move. My responsibilities required a whole different setup than what was available at the house Joyce had lived in while I was in Bosnia-Herzegovina.

In late 2003, following my year-long tour, I left command of SFOR and became the DCG, U.S. Army Europe. Like most units in the Army as well as other services, we were in full support of the Global War on Terror. I was responsible for preparing the Europe-based Army units that were involved in and participating in the War on Terror. During the same time, members of Congress were talking about force reduction in Europe and there were congressional hearings on force posture and forward stationing of forces outside of the U.S. Congress was getting involved with the stationing of troops overseas. We had the first Gulf War in the early 90s, so that is when the drawdown started.

When I was there in the mid-80s, there was a European force of 300,000 soldiers which was steadily being reduced following the fall of the Iron Curtain. All those activities were going on in Europe and as the DCG of U.S. Army Europe, I was in the middle of it all, training and equipping rotational units bound for the Mid-East and ensuring the units and families remaining in Europe were being cared for to the highest standard.

A Call from Secretary Rice

In early 2005, I received one of those unexpected, totally-out-of-the-blue calls. In my role as DCG, I had been out conducting a troop visit with an Army unit set to rotate to the War on Terror. My visit was meant to evaluate their training and preparation and assess the installation activities in place to take care of their families while the soldiers were deployed. All of this was essential to ensuring mission readiness.

My aide de camp said, "Sir you have a call from the

secretary."

Thinking it was my assistant back in the headquarters, I said, "No problem. I'll be back in the office shortly."

He said, "No sir. This is Secretary of State Rice from the States."

Wondering what this could be about, I answered, and a voice said, "General Ward, hold for Secretary Rice."

Condoleezza Rice came on the phone informing me there had been a conversation to discuss a special assignment for me and reporting directly to her: *"And we would like you to go to Israel for a short period of time, to be the U.S. Security Coordinator working with the Israelis and the Palestinian Authority."* Later I learned the 'we' in that conversation included her, Secretary of Defense Donald Rumsfeld, the National Security Advisor Steve Hadley, our U.S. Ambassador to Israel, Dan Kurtzer (Yes! The same diplomat who was our U.S. Ambassador to Egypt during my time as the Chief, Office of Military Cooperation in Cairo), and EUCOM Commander, Jim Jones.

Israel/Palestinian Authority Peace Process

The mission was to increase the confidence between the Israelis and the Palestinians and create space to move forward on the Peace Process. Part of that confidence had to be with the Israelis feeling and being more secure and more confident in a capable Palestinian Security Force. That would then give them some confidence that the Palestinians could be effective in helping to guarantee security there in Israel.

I was asked to come to DC for the orientation briefings and then go to Tel Aviv as quickly as possible. I said, "Well, okay. Yes ma'am." not quite knowing how that would evolve. I would find out later. Just as had been the case for my interview to be the executive officer to the Vice Chief of Staff, Army, here we are with another no-notice "hurry up and get to Washington mission."

I reserved a seat on a flight out of Germany back to DC for the next day. There was a series of meetings at the National Security Council, State Department and with the intelligence community. My focus was getting to know the mission and who I would be working with. As I'm about to leave the White House meetings, the National Security Advisor said, "General Ward, before you leave, we have one other stop."

A Meeting in the Oval Office

Steve Hadley escorted me around to the Oval Office. There, we saw President George W. Bush, who asked me to sit next to the fireplace with him. Steve Hadley joined Secretary Rumsfeld and Secretary Rice on the couch. Following a detailed discussion on what I was about to do, the President chuckled, "Now Kip, you didn't make any of these folks mad to get this job, did you!"

Although he was teasing, it was not lost on me that this assignment was something that had been considered and deliberated upon. In hindsight, part of the time I spent in Egypt probably came into play as they considered the role that the Egyptians played in Israeli Palestinian relations. I had built relationships with the Egyptians. I spent that year in Bosnia-Herzegovina and understood the interagency process and how that part of it worked. My experiences from those assignments were considered as beneficial to the dynamic for these five or six weeks in Israel, trying to establish confidence-building measures between the Palestinians and Israelis.

I was informed I'd be working directly for Secretary of State Condoleezza Rice, on a secondment from the Department of Defense, and would go to Israel for this period. While in DC participating in the various briefings and orientations, I familiarized myself with the principals and who all would be involved. I would be working with the U.S. Ambassador to Israel who had been our ambassador in Egypt. I had worked closely with Ambassador Dan Kurtzer when I was there as Chief of the Office of Security Cooperation. I remain in contact with him today.

When I was in Bosnia-Herzegovina, I worked closely with our embassy in Sarajevo and the various U.N. and European Union offices there. I'm sure all these factors came into play as I was selected for this job.

Building a Peace Process Team

I was given a timeline of when I would arrive in Israel. Flying into Tel Aviv, I would link up with the American Embassy and get another security detail assigned. This time instead of U.S. soldiers, these were U.S. Diplomatic Security individuals, responsible for the security of U.S. diplomats. There was an allowance that if I wanted someone from one of my previous security assignments, I could have them. There was an individual who had worked for me in Bosnia-Herzegovina as one of my security chiefs, Agent Mark Sills. I asked that he go with me, at least initially. That was great support, because knowing these women and men is important. Although the majority of my security detail were from Diplomatic Security Services, we had not had time to build a relationship. My chief of security would be an American soldier from the protective services division of the military police corps, U.S. Army.

We all joined up there in Israel. I met with Dan Kurtzer, the U.S. ambassador in Israel. I had my mission and understood the other American players who were in Israel. All were working towards helping build this confidence in the Israeli-Palestinian Peace Process, so that it might build and lead to a durable result.

Jim Wolfensen was my counterpart on the economic side. I was the U.S. Security Coordinator while he was the U.S. Economic Coordinator. Going back to my days in Bosnia-Herzegovina I understood the value of working together. I knew that he and I needed to have a close working relationship. I made a point of getting to know him. The USAID (Agency for International Development) staff were also there working, and I wanted to understand what they were doing as well.

Critical is the point of how we work together, to synchronize and harmonize actions. All those players were a part of the U.S. apparatus in Israel, trying to bring together a more coherent effort to help lead towards peace and stability between Israel and the Palestinians. I joined that and worked with it.

A team was being assembled to better understand the situation. Jim Bever was the USAID West Bank and Gaza Mission Director. Senior British development official, Emma Skye, from the UK equivalent of our USAID, was there providing international support and assistance in security and other matters. Emma had worked with our U.S. forces in Afghanistan earlier in her career.

U.S. foreign affairs officer Bill Taylor was there working on the economic side of things with Jim Wolfensohn, and I got to know him extremely well. Later, he and I would meet when I was the Deputy Commander of U.S. European Command (EUCOM), and he, the U.S. ambassador to Ukraine. We had things in common. Bill was a former military officer having graduated from West Point. When he knew that I had taught in the Department of Social Sciences at West Point that helped draw us together. We have been together on occasion at various foreign policy events since that time.

Things in your life happen and later come together in an unforeseen mosaic. I went to this assignment in Israel from Heidelberg as the DCG, U.S. Army Europe. Given the newness of the position, there was no staff. I was charged with pulling together a small team. I had worked in Bosnia-Herzegovina for the SACEUR (Supreme Allied Commander-Europe) General Jim Jones, who was also the Commander of EUCOM, my higher headquarters in USAREUR. Subsequently, he had been informed of my assignment to Israel. So, I went to him and said, "Sir, I need some folks to support me."

Because Israel was then in the EUCOM area of responsibility, I asked him for senior officer support to augment a few junior officers from USAREUR. Ironically, I was able to get an Army Colonel who worked for me when I commanded the 25th Infantry

Division in Hawai'i. Army Colonel Bill Puttmann was working in NATO at the Supreme Headquarters Allied Powers Europe. I said to General Jones that I'd sure like to take Bill Puttmann with me to Israel as my Deputy and Chief of Staff. After some deliberation, because Bill was a key member of the SACEUR staff, General Jones said, "You got it."

That is how I began building this team. (I was very appreciative of the relationship General Jones and I had established during my tenure as COMSFOR.) I had a young aide de camp, Army Major Steve Michael, who had jumped into Afghanistan with the 173rd Airborne Brigade based in Vicenza, Italy. During one of my early visits to Vicenza, I was reintroduced to Steve. He had also been in one of my battalions when I commanded the 2nd Brigade in the 10th Mountain Division at Fort Drum. I had just asked Steve to be my aide when the USSC mission surfaced, and his family had not yet relocated to Stuttgart.

I said, "Steve, Buddy, I need you to go to Israel with me on this mission." That is how I put this team together. Steve Michael, Bill Puttmann, Emma Skye, Leslie Darling, Doug Wilderman, and Major Luckett from the U.S. Army Europe Staff helped form a small team that would then go with me to Israel.

Within a matter of days, our team arrived in Tel Aviv on our mission to help with these confidence-building measures. Meetings with Prime Minister of Israel Ariel Sharon and other key Israeli defense and security officials were arranged. On many occasions, I met with the President of the Palestinian Authority, Mahmoud Abbas and the Palestinian Authority Security Chief, there in Ramallah in the West Bank; and I traveled to the Gaza Strip on several occasions, the only American official permitted to do so. The initial thought was that Israel would be a five-week mission—it wound up lasting over 11 months. Suffice it to say, we worked deliberately and diligently in trying to engender a degree of support, cooperation, understanding and confidence between the two forces. We spent a lot of time working on these confidence-building measures.

Working with Both Sides

At the time there were not a lot of resources, equipment and training coming to the Palestinians, so we were using what we had the best we could. We worked with the Palestinian Authority and Israel Defense Forces to understand where they were. We had some success in doing that, but lacking resources, sustaining our efforts was difficult.

I traveled to the Gaza Strip and the West Bank from Jerusalem on a routine basis. I was the only American official authorized to go into the Gaza Strip. Because of the hostilities, Congress had passed a rule that no Americans were to be in the Gaza Strip. I let Secretary Rice know I couldn't do my mission without being able to go there. So, a special dispensation was created to allow me to go into the Gaza Strip.

We went with a full security detail, but in a very subdued manner. The Assistant Secretary of State for the Near East and Asia would time his trips there with mine thus allowing him to go with me. He made a trip or two into Gaza with me because that was the only way he could get to Gaza to get a first-hand account of the situation.

I worked closely with the USAID officials in Jerusalem and Palestinians in Gaza who were there working with USAID programs. I wanted to align my efforts with those of the rest of the government so we could be better coordinated and more effective in creating the conditions for mutual confidence, trust, and security between parties.

Whether working with Israeli defense officials in Tel Aviv or Jerusalem or with Palestinian security officials in the West Bank or Gaza, my approach was to encourage them to see the proverbial glass as half full, to look at things from a positive point of view. People would acknowledge me, knowing I didn't want to talk about negativity and what was missing. If we were to start off talking about all the negatives, then we would not get anywhere. I was there to listen, to talk about building upon things that exist, and moving forward. That became the way we operated.

When I left the mission, Prime Minister Ariel Sharon told Secretary Rice, "We hate to see General Ward go because he has not taken any sides and works with both of us evenly." She received the same description of my approach to the work from President Mahmoud Abbas. As the person I directly reported to in the State Department, she recognized my approach. I knew from my understanding of the environment that the Israelis understood the Palestinians and the Palestinians understood the Israelis. Both groups knew, if I said one thing to one side, the other side would know it. I was transparent. If I told one that I thought something they were doing was not as effective as it ought to be or should be or could be, I would tell the other as well.

That dynamic supported my positive dealing with both entities, even during the Israeli Disengagement from Gaza in September 2005. Secretary Rice had asked me to do all I could to ensure the departure of Israel from Gaza did not occur under fire. That took a complicated system of behind-the-scenes coordination with me as the conduit between Israeli and Palestinian security forces.

When I left Israel returning to Germany in December 2005, Prime Minister Sharon and President Abbas were aware of the positive dealings we'd had with both sides and wanted me to stay on. However, by now, I had been nominated and confirmed to receive my fourth star and become the Deputy EUCOM Commander. (Again, Marine General Jim Jones's handprint was all over that action signaling the importance of prior joint assignments for general/flag officers.)

I did my best to see things from the perspective of those I served; this increased my effectiveness in that environment. I continue to use that great example of how to make a difference. Approaching and understanding situations from a perspective other than your own is essential. It is not that you are obligated to respond based on those perspectives but actively listening and understanding then impacts how you accomplish your goals.

The relationships and the trust that we established with the Israel Defense Forces and the Palestinian Security Forces were effective. And the heads of the U.S. and international agencies and organizations we partnered with made similar comments regarding our engagement with them. That approach was successful and solidified the mission's limited success.

I had experienced these sentiments before. When I left Egypt as the Chief of the Office of Military/Security Cooperation (OMC), President Hosni Mubarak said I was the general that got all of his generals to see things in the same way and that his generals and intelligence chief had only positive things to say. (I had convinced the Egyptians to allow me to visit and photograph their full inventory of armored vehicles resulting in U.S. Congressional authorization to produce additional M1 Tanks in Egypt.)

In recognition of that success, Mubarak presented me with the Egyptian Legion of Merit, a significant award and becoming the first American, not the U.S. Combatant Commander of CENTCOM, to receive it. [CENTCOM is the U.S. Central Command, one of 11 combatant commands of the U.S. Department of Defense. Its area of responsibility includes the Middle East, Central Asia and parts of South Asia.]

Now it's happening again in Israel when Prime Minister Ariel Sharon and President of the Palestinian Authority Mahmoud Abbas both indicated that my work there had been honorable. They wanted to see me continue the mission there.

Israel was a rewarding time. In addition to the mission, being in the Holy Land and seeing the actual ground and places referenced in the Bible was a tremendous experience for us. Joyce visited me in Israel and together we saw sites in Jerusalem, Jericho, the Sea of Galilee, and the Garden of Gethsemane among other notable Biblical locations. At this point, I was in my third three-star general assignment, including Bosnia-Herzegovina and the Deputy Commanding General (DCG) of the U.S. Army

Europe (USAREUR); and now there was this almost year-long assignment in Israel.

When I left that job in Israel, a permanent billet (position) for my successor was authorized by DOD, as opposed to an existing lieutenant general being detailed from a job to take the position as had been my case, an already existing three-star billet. Thus, my replacement was promoted to take the position I held. But that also would create additional stability as the officer would be there for a two- or three-year assignment. By then I had been away from my job as USAREUR DCG for almost a year with three years in grade as a lieutenant general and I'd been informed of my nomination for my fourth star. I had been in the Army for 34 years and it had been an extraordinary career.

Shortly after returning from Israel in December 2005 and knowing I would be the next Deputy Commander of U.S. European Command (EUCOM), I began to prepare for the assignment. At the time, the Deputy Commander of EUCOM was a four-star billet. This was because the commander of the EUCOM was also the Supreme Allied Commander of Europe, SACEUR.

The deputy commander is engaged in joint force activities and deals with the U.S. component commands of EUCOM. As a combatant command, EUCOM includes all services: Army, Navy, Air Force, and Marines, and each has a headquarters element. Those commanders were also four-star generals and an admiral. The Deputy Commander of EUCOM was a four-star billet and for me that would be another interview by the Secretary of Defense, a nomination by the President, hopefully a confirmation by the U.S. Senate and a promotion.

Confirmed Four-Star General, at last!

I was nominated and later confirmed by the Senate for my fourth star. In May 2006, I was promoted to full general as the Deputy Commander of the United States European Command. Subsequently I learned that one of the reasons I left Israel when

I did was because that was in the works. When General Jim Jones agreed to support my going to Israel, he knew then that I would be a strong candidate to be the next Deputy Commander of EUCOM. He wanted that for me and to this day I am grateful to Jim Jones. I had been introduced to him when he was the military assistant to the Secretary of Defense. Secretary Bill Cohen and his wife came to Cairo on a visit within a week after Joyce and I had arrived in Cairo.

General Jones let me know the Secretary and Mrs. Cohen would be visiting so we had extensive coordination over the phone. I was the senior U.S. Defense official in Egypt and therefore had great responsibility in coordinating the visit. Joyce had a critical role as well in coordinating the itinerary for Mrs. Cohen with the Egyptians.

A couple of days before the scheduled visit, General Jones called to tell me that he would not be accompanying the Secretary and for me to handle everything. (My boss at CENTCOM, Marine General Tony Zinni who I'd served with in Somalia in 1993 told him not to worry.) The three-day visit occurred without a hitch. Secretary Cohen and Mrs. Cohen were pleased with all that occurred and all their objectives for the visit were met. This was relayed to General Jones and he called after the visit with a remote high five!

Subsequently, I worked for Jim Jones twice: once as the commander of the NATO Stabilization force in Bosnia-Herzegovina and when I became the USAREUR Deputy Commander. Now I would be going to Stuttgart, Germany as his deputy at EUCOM. He was the Commander of EUCOM and dual-hatted as the Commander, Supreme Headquarters Allied Powers Europe (SHAPE) in Brussels, Belgium. I would be the U.S. Army's fifth Black four-star general serving as Deputy Commander of United States European Command!

CHAPTER 6

Standing Up AFRICOM

In May 2006, as a four-star general, I became the deputy commander of United States European Command (EUCOM). The U.S. Army was officially established in 1775 and in 2025 it had existed for 250 years. I had become only the fifth Black officer to attain this rank in the U.S. Army. This was another Joint job for me, working with all services and civilians, including the interagency departments and agencies of the federal government.

In addition to the European continent, EUCOM's area of responsibility included portions of the continent of Africa. The structure as established in the Department of Defense was such that three different combatant commands conducted military activities within the 50+ countries in Africa and its island nations: U.S. European Command (EUCOM), U.S. Central Command (CENTCOM), and the, then, U.S. Pacific Command (PACOM).

Having been the deputy commander of the U.S. Army Europe just prior to my promotion and year in Israel, I was not new to EUCOM. I had traveled to sub-Saharan Africa. The U.S. Central Command had responsibility for activities in the Horn of Africa region. The Pacific Command had responsibility for actions in the island nations of Africa in the Indian Ocean.

Because of the inaccurate design of world maps, often people are confused about the size of the African continent. In reality, Africa is more than three times the size of the continental United States. At the time, there were 53 distinctly different countries in Africa, including its island nations. Therefore, three separate combatant commands were responsible for dealing with all of Africa. Because of this, things were not always as coordinated as they might have been. The Africans had established the African Union, their continental organization, similar to the European

Union. The Department of Defense's decision to establish a separate command for Africa began to gain traction. Obviously, its construct would be a bit different, and it would have a unique interagency flair to it.

The new combatant command was not readily received in Africa nor Washington, reasons mostly revolving around territorial issues that were threatening to some African countries and U.S. government agencies for myriad reasons. There were interagency concerns, and the Department of Defense is a big entity. Other government agencies saw it as a behemoth of an organization that could overshadow and take over their mission. That wasn't the case, but that is how it was portrayed in some circles.

On the other hand, even though the United States had never colonized Africa, the notion of a high command for Africa with large naval battle groups and battalions of soldiers and squadrons of air forces coming in was threatening. That was never the case. However, there was pushback from that as well. It took us a year to dispel the perceived threats and gain acceptance by most parties.

During the year leading up to the establishment of the command, these things were being socialized. There was pushback from many corners and there was no blueprint for how to create this brand-new command from whole cloth—from scratch!

The Senate confirmation reappointing me to the rank of 4-star general occurred in August 2007; we stood up United States Africa Command (AFRICOM) in October of 2007 as a sub-unified command of EUCOM. That term *stood up* or *stand up* is military lingo for formally activating and commissioning a unit, formation, or command structure. While AFRICOM was a separate command, it was sub-unified to EUCOM until it achieved full operational capability, one year away. So, we organized the command with the initial operating capability as a sub-unified command, pulling together a staff and incrementally taking on missions.

Although President Bush and Secretary Rumsfeld had made the decision to stand up AFRICOM, the three combatant commands that were already operating in Africa were not eager to give up their missions there or the people supporting the mission.

Combatant commands would be tapped to support the headquarters' staff requirements, with portions of their staffs being reassigned to AFRICOM. So, we went through those kinds of negotiations as to how many from EUCOM, how many from PACOM and how many from CENTCOM. There was some back and forth. Admiral Mike Mullen, as Chairman of the Joint Chiefs of Staff, had me involved in staffing the command, which meant I traveled back and forth to Washington, DC on a regular basis.

We Built a Team

Because the command's construct included representatives from our interagency in the headquarters of the command, there were many interagency meetings. (Setting up this command and how it all transpired could be a book in and of itself.) In what seemed like a flash, we had the newest geographic combatant command—AFRICOM—established and operating; we made important strides and positive things were happening.

AFRICOM's Chief of Staff, Air Force Major General Mike Snodgrass likened our activities to flying an airplane while it's being built. There was enough there to keep it airborne; however, enhancements were being made every day.

We needed to build a team. The senior NCO that I chose to be the senior enlisted leader would be exceptionally valuable in enhancing the stand-up and functioning of AFRICOM. Typically, these men and women are nominated and assigned by their service headquarters; but with these assignments, as the commander, I had the opportunity to request the individual.

When I commanded the 25th Infantry Division in Hawai'i,

Mark Ripka was my Division Command Sergeant Major (CSM). Mark was an outstanding noncommissioned officer. I had kept up with Mark and knew he was about to retire. Following our time together in Hawai'i, Mark became the Senior Enlisted Leader at Joint Forces Command in Norfolk, Virginia. I knew he understood the importance of relationships and how soldiering aligned with other activities to make a difference. Mark would be the perfect choice to serve as the command's senior enlisted leader, but would he be available? Joyce was also aware of Mark's excellence and while I was hesitant about asking him due to his personal situation, Joyce encouraged me to reach out and ask.

He was about to retire when I reached out to him; we talked, and he said yes. But knowing he and his wife were going to be settling down, I said, "No, Buddy. You can't tell me yes until you've spoken to your wife."

After talking with his wife, Deb, he called to say they made the decision. Deb told him, "Well, if you're going to be working with General Ward absolutely." So, Mark Ripka became AFRICOM's first senior enlisted leader.

Part of the construct included having a Deputy Commander for Civil-Military Activity. At that time, AFRICOM was different from any other combatant command. Normally there would be one deputy in the combatant command, a senior flag or general (as had been my case in EUCOM.) Because of the uniqueness of this command, we also wanted to have a civilian deputy who had decision-making authority right next to that deputy military commander. That would be a unique characteristic of the command.

Selecting the civilian with the requisite understanding of the military and civilian components would be very important. Some ambassadors and other senior interagency government officials had those experiences. I quickly determined that an ambassador who had been working with me in EUCOM would be perfect for the job, Ambassador Mary Carlin Yates. The State Department concurred, and she became AFRICOM's first civilian deputy alongside Vice Admiral Robert Moeller, AFRICOM's first military deputy. Bob had led the transition team.

The services nominated other personnel for key positions in operations, intelligence, training, human relations resources and logistics. By and large I had no problem with any of them, they were fine with me. I assume competency and will work with what I'm given. For the most part over my career that seems to have worked out. I hope that is a measure and reflection of the type of leader I was in building effective teams.

The command was stood up and AFRICOM began conducting operations immediately. To this day, launching AFRICOM as its inaugural commander remains the highlight of my career. I viewed every command assignment as great, but this was the culmination of a career of engaging wholeheartedly in whatever task or mission I was assigned. In each case, the act of doing my very best to do things well, taking the lessons and learning from my previous experiences was instrumental in standing up AFRICOM as the United States' newest geographic combatant command. Although a new command, with no playbook, my personal template was a familiar one.

This experience served as a model for important organizational changes that could be made in other combatant commands and that in fact happened. The headquarters was there in Stuttgart, Germany. Not surprising, there was a lot of back and forth about why the headquarters was in Stuttgart and not in Africa.

A lot of speculation surfaced about what would happen. My time working with the U.S. Agency for International Development (USAID) was important as they were a big part of AFRICOM. When I was in Israel, I worked closely with Jim Beaver, the Director of USAID in Israel. That gave me an opportunity to see the importance of their work, firsthand. Observing the work I did representing the military and the degree of cooperation between the Department of Defense (DOD),

USAID and the different entities we represented, Beaver told me he had never seen that kind of connectivity before—and he was a senior USAID official. At that time and later, I understood the importance of USAID and the work it does on behalf of the U.S. When I left Israel in December 2005, Beaver gave me a little gift with the inscription, "You are one of this team!"

I carried that work experience with Jim Beaver to my work in AFRICOM and it influenced how we set it up; it was to work closely with State and USAID. When we set up AFRICOM, USAID Acting Director Henrietta Fore and Department of State Assistant Secretary Jendaji Frazer were a part of the inauguration ceremony hosted by the Chairman of the Joint Chiefs of Staff and the Secretary of Defense. We wanted the ceremony to show that this command was truly an interagency command unlike other combatant commands. Command Sergeant Major Ripka and I unfurled the colors and we ceremonially launched AFRICOM. It was a moment unlike any other because it was truly a joint interagency command.

Constant Travel/Service and Joint Sessions

Launching AFRICOM wasn't the end of the journey. The work, although building upon the military engagement already in place, was just beginning in a new and exciting way. Headquartered there at Stuttgart on historic Kelley Barracks, I constantly traveled. For a couple of weeks out of every month I crisscrossed the African continent: from Morocco and Algeria, to Kenya, Uganda, Sierra Leone, Benin, Nigeria, Ghana, South Africa, Botswana, Cape Verde, Cote' d'Ivoire, Democratic Republic of the Congo, Rwanda, Namibia, Angola and more. When I wasn't in Africa or back in Germany, I was back in the U.S. attending service and joint sessions, making visits to The Hill, and attending other interagency and civic activities in support of AFRICOM missions.

Many places in Germany had been former garrisons and military facilities. Thus, our house in Stuttgart was previous quarters for the German Army who used as an officers' lodge. The U.S. occupant was the Stuttgart community commander following the war, around 1945-46.

Over time it evolved into the residence of the senior military commander in Stuttgart. So, when the 7th Corps came in, the 7th Corps commander lived there. When EUCOM was established, the EUCOM Deputy Commander lived there. The EUCOM commander, dual-hatted as the Supreme Allied Commander, Europe (SACEUR) lived in Belgium at the NATO headquarters. As the EUCOM Deputy Commander, I became the first and only Black commander to live there.

The house was only known by its address, with no other name, which was not the case for most other senior residences in Germany. It was a 15,000 sq ft house, located in the city of Stuttgart. Not being on a U.S. installation, it was its own little separate compound, adjacent to property occupied by the governor of Baden-Wurttemberg, a state in Germany. Having been taken by the German Army from a German dentist before World War I, it was maintained by the German Army through World War II. After World War II with the defeat of the Nazis, the Americans took over many of those former German facilities and installations, including 39 Richard Wagner Strasse.

Joyce and I wondered why it had never been named. I reached out to the command historian at Stuttgart and the German community leaders, with strong U.S. relationships including Marianne Zinser, who was prominent in the German-American club. We asked what name the house could have that would resonate with everyone, U.S. and German, and make sense. The story of Lucius Clay kept surfacing. Known for his administration of occupied Germany after World War II, General Clay was instrumental in beginning the reconstruction of post-war Germany.

The Clay Haus

Given all that General Clay had done, we named the residence

The Clay Haus/House. The Americans loved it, and the Germans loved it. After I left the command, the U.S. Army turned it back over to the Germans, largely due to the expense of maintaining it. The Germans honored the renaming and kept the house name, located next door to the governor of the state of Baden-Wurttemberg.

As the senior commander in Stuttgart, maintaining a relationship with the German community was important. They were our hosts, and we spent a lot of time cultivating relationships with the mayor, local officials, and other Baden-Wurttemberg community members. Joyce and I remain in contact with former Stuttgart Lord Mayor Wolfgang Schuster and his wife, Stephanie

Like any relationship, you must work at it. We made sure to be present during holiday celebrations. The Clay Haus sponsored many Christmas gatherings, a tradition that began long before my time. Still the Senior U.S. Commander in Stuttgart and as Commander of AFRICOM, I continued the tradition. These were pleasant times and instrumental in maintaining supportive relationships. We hosted ambassadors, defense chiefs, U.S. officials, other dignitaries and our friends at the Clay Haus.

> ***Foxhole Lesson: Effective leadership requires cultivating meaningful relationships with all stakeholders – families, civilians, community members, foreign nationals, etc. – not just those in your chain of command. It all impacts readiness.***

We had arrived in Stuttgart, Germany to EUCOM headquarters when I was the Deputy Commander and remained after standing up AFRICOM there. We lived in that house for five years, the longest Joyce and I had ever lived in any place in our military career. Building the command meant meeting with various stakeholders: the interagency, as well as nongovernmental stakeholders—explaining the concept of this command and how it was a unique opportunity. In addition to doing the mission, I conducted a lot of strategic engagements because as we were building the command one of our tenets was: *Nothing is going to stop. As we build this command, the work goes on!* Our Department of Defense activities, our exercises, our military activities and our partnerships and relations—all of it was still going on.

Origins of the 3Ds (Defense-Diplomacy-Development)

It was not as if the Department of Defense had not already been operating in Africa. It was being done through EUCOM, CENTCOM and PACOM. We didn't want to build up a new command and have all that work suddenly cease. We needed to continue the mission. The work in Djibouti, in the Horn of Africa—our largest and only continental location with troops stationed at a semi-permanent location—was a primary example. All work needed to continue; however, we needed to find ways to add value. We didn't want our military engagement and operations to stop. We just wanted to make activities more coherent, cohesive, and synchronized and harmonized with other U.S. government actions.

We were building this command piece by piece, simultaneously continuing ongoing activities; in fact, we were making them better. In conducting our mission analysis there at the Combined Joint Task Force-Horn of Africa (CJTF-HOA) in 2007, the team and I coined the *3Ds* Term (Defense, Diplomacy, Development) that later proliferated throughout our government. This was an outgrowth of AFRICOM's purposeful and focused effort to better harmonize our defense efforts with the diplomatic role of the State Department and USAID's developmental programs.

The staff, the team, and the service components were a part of AFRICOM. We all knew it was not just me; it was an effective and energizing team effort. As people came on board, the good news was that we were able to utilize a lot of what existed in the other component commands to carry on. There was good cohesion as we transferred responsibility and respect for the gifts and skills each person brought to the team. In some cases, many of the same people were involved, having previously been assigned to another command. That made things more seamless than they might otherwise be.

I paid a lot of attention to my defense attachés (DATTS) and security cooperation officers (SCO). This is a group of men and women from all services—Army, Navy, Air Force, Marines—who conduct DOD activities and operated in our embassies around the world. (I had been one). People would say, "Well, you don't have any troops assigned to AFRICOM." I would say, "Yes, we do! They are already there." Those attachés, security cooperation officers, and defense officials *were* already there and a few years before, I was one of them in Egypt working for CENTCOM. I knew that piece of it well. They were my frontline eyes and ears and we were able to move through the Department of Defense, a change in the structure of the relationship.

At the time in an embassy, there were the defense attachés and the security cooperation officers; and they worked for two different headquarters. That didn't make sense. Now, we were able to realign so that both communicated, connected with, and worked for the combatant commander. Of course, they maintained close relationships with the country team and ambassador and they were responsive to their previous commands, but they worked for the combatant commander.

That model was adopted by the DOD and transported to other combatant commands. The senior defense official (SDO) was the overall person in the embassy from a military point of view with the responsibility, either the defense attaché, or in some cases, the security cooperation officer. But they all worked for and reported to the combatant commander. That connectivity and cooperation was an outgrowth of this arrangement at AFRICOM.

I knew the value of the work done by these defense attachés (DATTs) and the security cooperation officers and how important they were to our partner nations. They were integral in making the policies that we put in place on the ground work effectively. They spend a lot of time in operations, visiting countries, talking to defense officials and talking to host nation officials, be they presidents or prime ministers, getting to know them. They are the true experts in these partner nations.

A lot of my time was spent coming back to the States, visiting European capitals, and working with our European partners. It needs to be remembered that most of these African nations, having been colonized, were working with European nations like Portugal, France, Belgium, England and Spain; so, we had to make visits there as well, synchronizing and coordinating. We wanted to make sure that what we were doing would work. This helped us work better.

In testimony to the value of our International Military Education and Training (IMET) program, I quickly discovered that in France, the Chief of Defense Staff had been my classmate at Command and General Staff College at Fort Leavenworth in 1982. Our reconnecting reinforced the importance of relationships and the value of that program.

As a commander, you don't foster relationships from sitting behind a desk or at a computer. Meeting people face to face, listening to them, and sharing ideas brings the opportunity to learn from them. Spending time traveling and connecting paid off and strengthened those relationships, thus resulting in more effective military programs.

> ***Foxhole Lesson: Whenever possible, foster relationships; they matter. Get out of your foxhole, go down range and look back at it from another point of view. You'll see it differently and make better decisions.***

We Did That!!

Typically, my team(s) would be working out of embassies, where I would visit them. For example, all services might be working out of the U.S. embassy in that country, including Army, Navy, Air Force, and Marine personnel. Although their individual services provide their support, they are working there together. I would go there and on many of these trips my wife would accompany me. Her travel was always authorized, as it supported the mission and enhanced the wellbeing of our military families on the continent.

Joyce is one of those people who connects the dots quickly. After my meetings with the officials, as we traveled back, she would say something like, "Did you know that if there is a pregnancy in the embassy, the Air Force treats it differently than the Army and the Navy?"

I didn't know that, but the service members knew it. Members in the same location were getting different treatment depending on their Service. We would work on those kinds of policies in order to enhance the quality of life for people working there and we were doing that constantly. Folks understood it, knew it, and really appreciated the efforts we were making to get the mission done; and efforts to take care of our people. That is critical and reminded me of my motto in brigade command — "Mission First, People Always."

Service members and spouses who were there in the embassies were invited to the headquarters for synchronization conferences to make sure we all played from the same sheet of music. Inquiries would be made about what they needed and what they were doing so that they would know the staff is supporting them. These communications made a difference in moving ahead and many continue today in substantial ways that support the mission. These relationships grow within themselves, adding to the richness of the experience. Like anything else, criticisms surface—that's okay. We move forward and continue to improve the foxhole. Looking back, standing up AFRICOM in Stuttgart and building that command was a big deal and a great time.

Previously, combatant commands were taken from some existing headquarters and converted to a new organization, with a foundation, a structure, and an organizational entity already in place. This was not the case with AFRICOM. Together we had created an organization with a mission to be responsible for conducting all Department of Defense activities on the continent of Africa.

We built a team from the ground up into being a relevant and meaningful organization. It was important to me to understand the positions of those who were critical of an organizational construct that included these various interagency partners and leery of a military construct that was misunderstood by some and feared to be an occupying force.

These myriad perspectives existed amongst the U.S. agencies and our allies and friends, but also from our African partners and friends who would become the ones we were there to work with and support. We had to deal with these opposing and negative reactions to the standup of the command, and we had to build a team that could do that. In addition, we had to do it in such a way that none of the ongoing activities were impacted. In some cases, as a new command is created, some existing activities are negatively impacted.

I followed a well-known principle in how we would move forward: the Hippocratic Oath. Specifically, our purpose became to add value, do no harm, and enhance and grow programs, but at a minimum to do no harm to the existing activities of our government.

The philosophy that those in position are responsible and competent enough is something I've accepted throughout my career. It's not my philosophy to bring in different people to replace folks already doing a job. In the case of AFRICOM, we had to assimilate staff from three different commands, plus include others already assigned and doing work across the continent of Africa and the island nations: from European Command, Central Command and Pacific Command.

Certain personnel would be reassigned to AFRICOM from those commands, most were working on Africa-related issues. With the pushback, those numbers wouldn't be enough. Other staff were being assigned through the various service-assignment processes. AFRICOM as a Joint Command would include Army, Navy, Air Force, Marine, Coast Guard, Special Operations and civilians. Therefore, part of the normal rotation and assignment processes would route individuals to AFRICOM.

Services did offer selected key staff members, and I interviewed several of them prior to assignment. Having already selected the senior enlisted leader (SEL) who would be coming to the command, I certainly understood the value of noncommissioned officers and how important they were in getting the job done and mission accomplishment. They were critical to having cohesive teams and making a difference. In a military organization, the SEL is unquestionably valuable to that construct.

As a career infantryman, I worked with NCOs in all my assignments and in my college days. NCOs have always been important for what they bring to organizations and units for their expertise, their perspective, their understanding of how to make things happen. Civilians from across the government came to the command as the team was being built.

There were instances where multiple services would nominate a candidate for a critical position. The Director of Intelligence and Knowledge Development was one such position. I interviewed and hired Terrence Ford. Terry was a retired Army intelligence senior officer with an extensive follow-on career as a civilian in one of our intelligence agencies. He brought a unique understanding of military and civilian intelligence expertise as we sought to revolutionize typical military intelligence activities in favor of intelligence and knowledge development. (We needed to have cultural understanding to be effective in our area of operation). I requested Ms. Heide Grant for the resources chief position because of her stellar reputation as a competent and expert resource manager.

By and large the team was built because of who was in place. I took who was given to me. Until circumstances prove otherwise, my assumption is that you can do your job and be an effective teammate. I worked to reinforce that and that is what we did. So, as we built the AFRICOM team, it was a team that worked well together. Because there was no blueprint, we had to do and deal with a lot of things on the fly. We had to establish policies and procedures to get things done and move forward, all while not having a lapse in the activities that were already happening.

Partner Nations

Before AFRICOM, several military-to-military exercises and training activities with partner nations on the continent of Africa were already underway. I looked forward to enhancing these activities. Because I had been in one of those commands with EUCOM, I was aware of the various associations, relationships, and places on the continent where the other two commands were operating.

I wanted to build on the relationships while not having a negative impact on what our partners had come to expect. The goal was simple: to work with African partners so that they were better empowered and more able to provide for their own security and maintain a secure environment. I also understood that more than the security forces had a role in this undertaking. These goals were aligned with our mission to protect the United States and its citizens, at home and abroad.

Our interests were best served as a nation when peace and stability existed across the globe, including Africa. This fostered our ability to engage in commercial activities, have access to markets and resources and reduce, if not eliminate, the threat from terrorists. If nations could do their part in ensuring that type of security, then that was in our best interest.

The mission of AFRICOM was to work with these nations to help them increase their capacity to provide for their own security in the regions and areas of their sovereign territory. Once we were able to get that message to them in a substantial way, it allayed these fears of a dominating U.S. presence or of a force that would come in to take over their territory or infringe upon their sovereignty. That really mattered.

Additionally, the Department of Defense's arm for conducting security activities in Africa enabled other agencies and departments of the U.S. government to perform their mission better. This included the U.S. Department of Agriculture, where. food security was a big issue, so working with African nations in agricultural and commercial activities was essential. American businesses were thus better positioned to work on the continent. The U.S. Department of Commerce was a player and to be sure the Department of State. Working with our diplomatic missions around the world and working with the Department of State on American goals had a role. The USAID was also a part of this construct, as were other federal agencies (those three letter agencies.)

Work also included getting a handle on the intelligence picture, terrorism, counterterrorism: all of that, in conjunction with a plethora of U.S. government departments and agencies, also engaged on the continent.

As the nation's newest command, AFRICOM looked at its role not as competing with these other efforts but working in conjunction and coordination with them; synchronizing with them so that collectively our total efforts made a difference in supporting U.S. national policy interests and protecting American citizens abroad and securing America at home.

To do that building, this team included personnel who understood the missions of these other agencies and departments. The best way to do that was to have these personnel with us day to day as we observed, analyzed, and noted what they were doing and vice versa. There was some hesitancy because the DOD carried the big budget, and a prevailing concern was that the DOD was coming in to take over the existing missions and roles.

That was not the case, but I and other senior members of my team developed relationships with departments and agencies in the States, embassies on the continent, and country teams so that the understanding was: *No. We are not doing your jobs, and we don't want to do your jobs. We want to understand how you operate so that what we do together creates a more favorable result than independent actions that may be disjointed and create confusion and conflict.*

As we worked to make that the case, the team increasingly realized and accepted that working together was the way to go and made a difference for all of us. I am so pleased that after a shaky start, due to early doubts and the lack of understanding, the AFRICOM model proved to be valuable in engaging with partners.

We were able to get the support from other federal government departments and agencies. As a work in progress, we didn't get all the personnel we wanted when I was in command. However, the process continued and continues. Many of the processes, procedures and operating guidelines we put in place then are still being practiced today; they are sound and foundational in making progress across our various mission areas.

3Ds: Diplomacy-Development-Defense

It was during this period, 2007-2009, that we at AFRICOM introduced the 3Ds to our government: that Diplomacy-Development-Defense were the legs of a three-legged stool that needed to work together in producing a more stable environment. It kept with our interagency focus of bringing more cohesion, harmonization and synchronization to our government activities in our area of responsibility.

We understood the role of our security and military

programs and the importance of developmental activities that helped the people thrive where they live. With this was the need for effective diplomacy and governance to secure stability once peace is established. AFRICOM doesn't get the credit for this. I spoke to it during my engagements with Congress and in my presentations to governmental partners and think tanks. A few years later more and more officials began using the term as a foundational construct in our government.

Now when it came to the Africans, there were those who were resistant to the idea of AFRICOM because of the mistaken notion that it would be an occupying force, with large garrisons of soldiers and squadrons of planes. Although the U.S. had not been a colonizing force in Africa, the negative connotations of colonial rule were ever present.

In presentations and conversations, I was very deliberate in dispelling the notion of an occupying force. As I traveled across the continent, it was important for me to talk to public officials and leaders, civic groups, young people and the general populations to hear what they thought and to understand and talk about their perceptions. Those discussions helped to allay fears that AFRICOM would be a large command of garrisoned forces occupying their sovereign land. And with those collaborative interactions and our execution on the ground, soon those misconceptions and myths faded.

> ***Foxhole Lesson: Take people's concerns seriously; they are usually grounded in some reality. Do your best to allay fears, and let your actions do the rest.***

Over time, as we began to operate as an independent and fully operational command, they saw that our actions were aligned with our words. They saw that we did what we said in operating with our partners. To be sure, we had exercises and troop rotations, but there was no permanent stationing.

American Forces were there to come in and work with our military friends and partners, and they left the area better than before. For example, if a road or other infrastructure was needed for an exercise, we ensured it could serve useful purposes for the civilian community after the exercise.

With our U.S. government partners, we worked on projects and programs that made a difference to the African people, with positive impacts and outcomes. We worked collaboratively with the U.S. Department of State and USAID and sought to understand and work with various non-governmental organizations. It was important to understand the work NGOs were doing on the continent so that we didn't impact their efforts negatively. It was the same with our interagency partners.

The good news is that over time our words, actions and follow-up were reinforced. As the results of our conversations and our actions were more and more widely known, our value was realized and the command was increasingly accepted and respected. AFRICOM's existence was solid so that by the time I relinquished command— three and a half years later—almost all negativities had dissipated. And just as with the 3Ds theme, the command's foundational precepts remain in force today. The command was seen as a force for good and making a difference in creating a more stable environment. It goes back to AFRICOM's original purpose: We are not there to provide your security. We are there to assist you as you provide security for your own people.

This purpose reminded me of Nelson Mandela's words on an exhibit at his museum in South Africa, "We want to provide for our own security, but with the help of our friends." Our premise was that AFRICOM was there to support and enhance the efforts of African Nations and their security forces to provide their own security. The U.S. Africa Command was not there to perform duties as an occupying security force provider. Our focus was on *sustained security engagement* through military-to-military engagement, exercises, professional training, and kentic-engagement as required.

Creating Stable Societies Requires 3Ds!

In recognition of the 3Ds theme of Defense-Development-Diplomacy, I emphasized the importance of sustained developmental (economic/social) and sustained diplomatic engagement alongside sustained security engagement. Creating more stable societies requires all three to be working, with consistency.

By 2010, Secretary of Defense Robert Gates had implored all combatant commanders (COCOM) to highlight the importance of our developmental and diplomatic programs. The other COCOM commanders began doing what I had already been doing. All of us realized this was the long-term solution and less expensive than a sole focus on the defense leg.

As we supported and worked with our African partners and friends, it increased their capacity to provide for their own security. When government and community leaders were more capable, more professional, and security forces were operating in ways where their populations could see them as protectors and not oppressors, they could see how those behavioral changes benefited and supported their interests. This led to their increased stability and thus was in the interests of the United States.

We did what we did because that helped guarantee the safety and security of Americans at home and abroad. These nations were less susceptible to the influence of bad actors and terrorists seeking to exploit them. In the end, it was about how to make a safer and more stable environment at home and abroad. And that is what we did.

Looking back at standing up that command, among all the stories, what I saw as critically important was building relationships, being a listening and learning organization, and reinforcing successes. If something wasn't working the way it needed to work, then we would take the feedback from our partners and friends and incorporate it into our actions to get a better outcome. That our partners had buy-in to the result was equally important. Evaluating our processes—especially after executing and completing an assignment—became as important as the planning and preparation.

Our partners and friends came to know that and there was a healthy respect for building *partner relationships/partnering for peace*. As we partnered for peace, it wasn't a: *We're the United States/We know everything/Do it our way or it's the highway* situation. We actively listened to the input. The partners working alongside us also had ownership and responsibility in the outcome. That was the intended result.

Of course, all of that was a work in progress. It wasn't something that you would do and then achieve success and then leave it alone. Like many relationships you intentionally work at maintaining them. For me and staff members like my senior enlisted leader, the command sergeant major and other senior members of my staff, the two deputies and the directors of the various departments—we all, with strong commitment, supported building and maintaining those relationships. This included having relationships with and working with the embassies and with military personnel (the attachés and offices of security cooperation) assigned to embassies.

I had a direct role in redefining and redesigning how those military personnel in the embassies worked within a combatant command. I wanted to ensure that all military personnel who worked in the embassy as part of the ambassador's country team, would have no doubt that they would also be a direct link back to my headquarters. This was beyond the senior defense official who worked directly for the Combatant Command.

This extended to all service members, be they Army, Air Force, Marine, or Navy. The service members and their families stationed on the continent were my charges and their well-being was my responsibility. My staff and I had a mission to ensure they were cared for and we did just that, in ways they realized.

My having direct rating authority over those persons helped

and when people challenged AFRICOM's personnel needs, asking, "Well, you don't have any forces in your command!" I would show them that we had forces all over that theater, assigned in the form of defense attachés (DATTs) and the Offices of Security Cooperation, and that the personnel assigned in those offices worked day in and day out with our partner nations, pursuing their military advancement requirements and our national interests.

On the continent we made sure African ambassadors and colleagues were aligned with our overall U.S. mission and goals. I coordinated with U.S. ambassadors and Chiefs of Mission, so that they were fully involved in what we did in their countries and on the continent. For instance, they had the ability to provide input, in letting us know that some procedure we might be planning could be disruptive to some local circumstance or situation. Then together we would figure out how to work out the complication.

We built relationships and they weren't one-offs or one-time interactions; they were continual relationships, day-in/day-out connections to ensure cohesion, synchronization, cooperation and coordination. In terms of how we operated, the interests of America were better served. Many other combatant commands and departments adopted that model, including having a Deputy Commander for Civil/Military Activities (later military engagement).

Programs for the Population

In most military organizations, there is a commander and a deputy commander directly under that commander. In the case of AFRICOM, we built a design that included one military deputy commander, and a civilian deputy to the commander. The civilian deputy was a career ambassador from the Department of State.

We paid attention to the alignment of the military activities with diplomatic efforts and economic efforts. To that end we also asked for a senior development officer from USAID to be a part of the team. We also worked with other U.S. government departments and agencies to have other senior civilian personnel assigned to the command. By the time I changed command, representatives from Treasury, Commerce, State, Agriculture, USAID, and the interagency were working in AFRICOM.

Those civilians weren't just there to be subservient to or simply to acknowledge what we did. They were there to be instructive so that we knew what they were doing, and our actions meshed with the activities of those other stakeholders active on the continent.

This model created that space for them. This kind of communication had not existed in the Department of Defense before. Because of how we were able to implement it and show its value, it has now been adopted across other combatant commands. I'm thankful that we looked to do business in a different, more effective, efficient and productive way; it was in the interest of pursuing our national interests in the United States.

There were detractors and those who thought creating a new command was intruding in their territory. They saw this new way of doing business as a disruption to their work; but we just had to move forward. And moving forward caused us to spend a lot of time out of our headquarters, traveling and engaging with potential stakeholders in Europe.

Given the European's earlier colonization of the continent of Africa, they still had a stake there. So, our coordination included connecting with European friends, including the British, French, Belgians, Italians, and Portuguese. We wanted to understand what these stakeholders were doing to understand the implications for our programs. Including all stakeholders added value to the total effort.

We did a lot of relationship-building in Europe as well as on the continent of Africa, and we developed relationships and partnerships with various influencers in the United States. This included meetings with think tanks and the service schools. We needed to have these discussions so they could understand the new command, would be less fearful of what was going on, and could buy into it. The good news was that most who heard about AFRICOM's purpose, progress, and value appreciated and supported the command. For three and a half years, our team produced all of that. It did not happen overnight, nor could it ever be taken for granted.

Through all of this we had a lot going on caring for families and spent many hours working to perfect the standards of family readiness and family support, and it made a difference. On the continent, my defense attachés and security cooperation officers were from all the military services: Army, Marine Corps, Navy, and Air Force, each being supported by their respective service.

This created some inconsistencies inside of U.S. embassies in various countries. With one service treating a family situation one way and another service handling it differently, while both served in the same location, the new command solved this disparity. We learned how dysfunctional it was to the team right there on the ground when they saw different treatment being given to the other, when both were doing the same mission for the United States of America.

As I traveled, Joyce and I spent time identifying those issues. She facilitated meetings with family members on the continent and participated in discussions with the various civic organizations in different African countries, including schools, orphanages and medical facilities. Her conversations and relationships included women's empowerment issues because all of that was in support of our mission activities; they added value to those we interacted with and that was an important part of the process.

Enhancing the quality of life and the well-being of military families on the continent was a critical objective. We needed to address various issues impacting the education of American children, working to give them the same enriching childhood experiences that happen for military children elsewhere. This included incorporating activities from Sesame Street in New York, and coordinating the delivery of books and other educational materials for military children and young readers in Africa.

We developed or implemented well-being programs for family members. The U.S. military had long discovered that family enhancements added to the readiness of the force and effectiveness of the mission. However, a lot of people just didn't understand; even some people in my headquarters wondered: *Why is this combatant command paying attention to how pregnant women are treated?*

Our response: *Because that pregnant woman's spouse, who is a service member, is concerned about it! That service member cannot perform assignments wondering and worrying about a pregnant partner*. Those were the issues and challenges impacting readiness and effectiveness that we spent time trying to address to make a positive difference, and we did.

In terms of the African population, we also knew the role that women played in most of these countries in stabilizing the society. Their entrepreneurship and small business efforts were instrumental in maintaining stability and helping make these populations less vulnerable to terrorists seeking to exploit them. It was easy for me to see that when a mother in an African family was able to feed and clothe her children and send them to school, that meant that she would also be there keeping her vulnerable sons from the influence of negative actors and terrorists. When young people know where their next meal is coming from and are afforded the set of options that comes with an education, they are also less susceptible to the allure of nefarious organizations and the promises they make to vulnerable populations.

That example may seem insignificant, but it's universal. The nationality of the kid does not matter. Kids in Bosnia, Israel (including the Gaza Strip), Somalia or America share universal values. For students at George Washington University, I did a lecture about this issue as it relates to stabilizing communities and the role of education. If a family can keep its young boys (and yes—it's mostly boys) fed, occupied, and relatively healthy, the propensity for them to go in negative directions is decreased, perhaps even as much as 98%.

Societal Needs and Security—Horizon of Hope

For sure, a major part of establishing security includes actions taken to improve the quality of life for its citizens. This includes economic and educational opportunity, access to healthcare, food security, and other dimensions of well-being. It is as true in Abuja, Luanda, Marrakesh or Kigali as it is in any American locale, the same dynamic. Policies, procedures and military activities should be aligned simultaneously with societal needs. We made that a part of this military organization and folks were having to come to grips with the construct. Now the connection between military success and civilian well-being is certainly more widely understood and today programs exist that directly address that.

Providing programs for the basic needs of the population is an essential element in keeping people satisfied and optimistic about the future. Optimism comes because basic needs for food, safety, education, health care, and resources to provide for the needs of one's family have been met to some degree. After enduring national challenges or tragedies, people must have the *horizon of hope* I referred to earlier—this hope for a better day rooted in opportunity and potential is a stabilizing factor in societies. When there is hope, people are less susceptible to bad actors and engaging in criminal actions, terrorism and other nefarious influences.

When I talked about those linkages, sometimes folks looked at me as if I had horns coming out of my head. Why? Because I thought those issues were an important part of the overall dynamic. Even today as I work with organizations and councils, I talk about the importance of creating the *horizon of hope* to counter bad actors and keep folks satisfied where they are. It is seen and understood.

These priorities and actions complemented our defined military missions like counterterrorism and establishing peace; of course we did that too. But that wasn't all that this command was about, and I wanted to ensure that folks saw the totality and completeness of AFRICOM and what it meant.

To this day I encounter people at various activities and events who let me know that what we established is still working and thank me for emphasizing it so much.

In each of my earlier assignments, I went into an existing organization where I took what was there and did my best to improve the foxhole. That meant to make it better every day and cause those who are a part of the team to make sure that their part of the foxhole was also being improved every day. But in the case of AFRICOM there was no existing foxhole. We had to create the foxhole and work every day (and think hard on it at night) to make it better as we were building it.

My Air Force Chief of Staff always used an example from an Air Force pilot's perspective. His point was: *"We are building this airplane as we fly it."* He illustrated an airplane flying with only a skeleton, but still flying. As the builders were trying to strap on the plane's sides and add equipment, the plane wasn't on the ground sitting still! It was flying while being built! That is what we were doing with AFRICOM. It was in full motion while simultaneously being built, because there were missions already underway that had been initiated under other commands that we took over and improved upon.

Clearly the command has continued to evolve. We've gone from the directorates back to J-Codes (i.e., from human resources, intelligence and knowledge development operations, logistics, planning, and so on back to J1-J9) just in the nomenclature because of the familiarity, but the functions and things that we were doing have remained the same.

Let's improve the foxhole in showing that we are considering the perspectives of others, listening to our stakeholders, and then incorporating good ideas that make sense into what we do. It is not that we must do everything that gets presented to us, but we need to take it into account and evaluate and assess all options. When we do that, we are a better organization.

> ***Foxhole Lesson: Keep improving your foxhole.***

Similarly, when organizations become more diverse, they become better because organizers receive the input about concepts they wouldn't have considered otherwise. If that input makes a difference in improving and enhancing the effort, the organization is better off for having it. You don't get that if you don't have a diverse organization.

We valued diversity and understanding. When we built AFRICOM, I wanted people who understood the continent of Africa. Most of the people who were being assigned to the command had not served in Africa. Some of them may have had some brief association in some way, but most had no real or educated perspective.

So how do we go to a continent and work to make things better when we don't understand it? We acknowledge that deficit and then bring in folks who have an understanding to tell us and teach us and inform us. With their help, when we go, we have a better appreciation of what to expect; we gain an understanding of and a respect for that location, and we learn the do's and don'ts.

For our cultural education, I asked for some cultural anthropologists. That request threw some for a loop. They questioned, "Are you out of your mind? You are a military organization. What has culture got to do with your responsibilities?"

My response: *Well yes! I need someone to help my team and me understand this continent. And the cultural needs of one country don't fit the needs of another. These are not one-size-fits-all situations.*

I couldn't be too hard on them—most folks don't comprehend how large Africa is and it's difficult for people who speak only one language to digest. At the time, on the continent there were over 800 languages and dialects spoken in 53 countries, 5 regions (and some subregions), and all were discrete and different!

Learning one thing in Southern Africa doesn't mean that same thing will fly in Central or North Africa or East or West Africa. We needed cultural anthropologists to help us understand the rules and expectations of the culture(s) and societies so that our presence and behaviors did nothing that could be contrary to the results we wanted to achieve.

That was highlighted on my tour in Egypt as the Chief of the Office of Military/Security Cooperation. I had my cultural training from the State Department prior to going to Egypt. The State Department does that for its people. Well, us military folks, we didn't do that. But AFRICOM did! And it was a good thing. It was for understanding the basics of the society in terms of customs and traditions so that we weren't out of place or perceived as the proverbial ugly Americans. Even stronger, it was just plain human courtesy to respect a person, especially on the road to relationship-building.

All my commands were great. But clearly the achievement of being the inaugural commander of the U.S. Africa Command is a personal summit. That I had a major part in establishing it, watching it progress, creating its legitimacy, and that it impacts and adds value to so many around the world, including our friends on the continent, allies, and other players brings me a sense of great joy. Importantly, we developed lasting friendships in our government too. That continuous ripple effect affirms what we are doing and where we are.

Now almost two decades later, AFRICOM is still going strong. The base that we established in 2007 is still seen as a very adequate base for building upon to make a difference.

AFRICOM remains a crowning assignment for me and is one that I am so blessed and proud to have been able to do. I am thankful for all that the team was able to achieve in making a difference for the security of Americans, at home and abroad; and in enhancing the welfare of our friends and partners on the continent of Africa and its island nations.

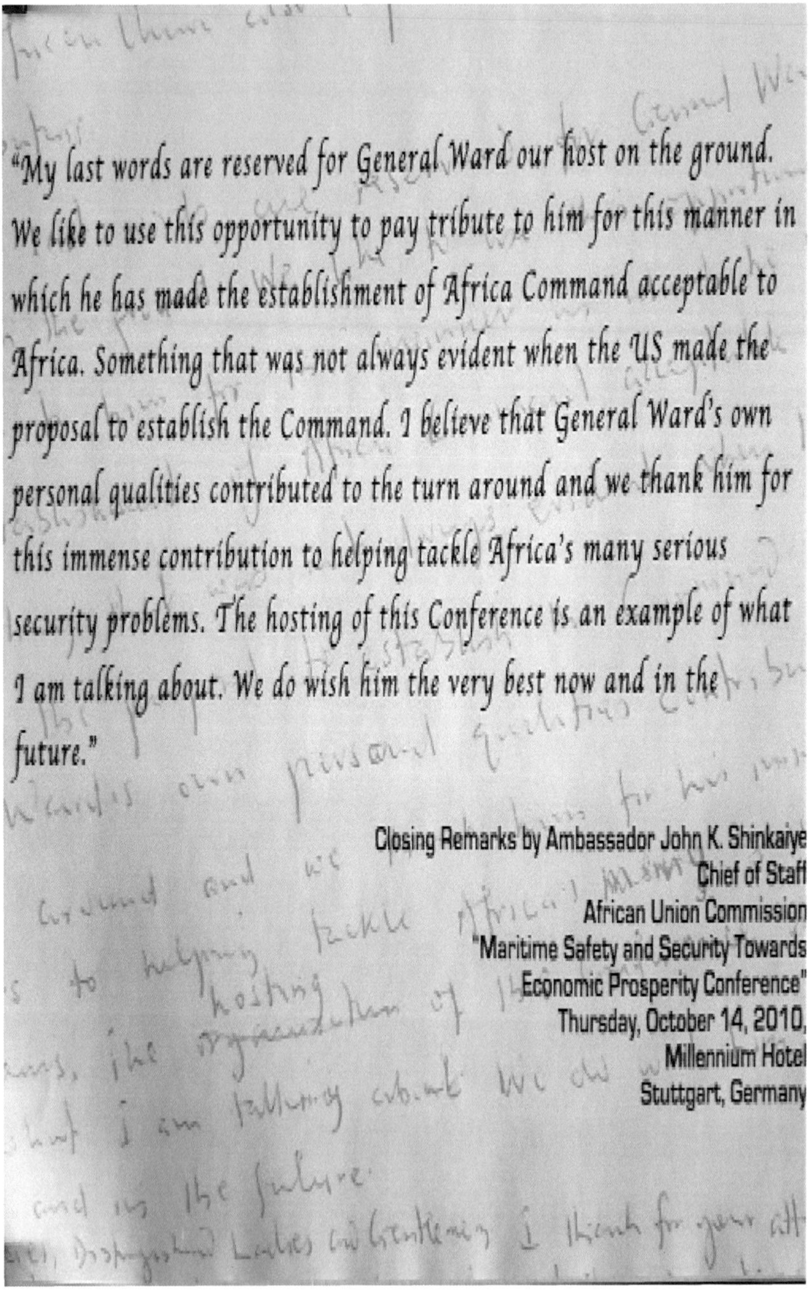

"My last words are reserved for General Ward our host on the ground. We like to use this opportunity to pay tribute to him for this manner in which he has made the establishment of Africa Command acceptable to Africa. Something that was not always evident when the US made the proposal to establish the Command. I believe that General Ward's own personal qualities contributed to the turn around and we thank him for this immense contribution to helping tackle Africa's many serious security problems. The hosting of this Conference is an example of what I am talking about. We do wish him the very best now and in the future."

Closing Remarks by Ambassador John K. Shinkaiye
Chief of Staff
African Union Commission
"Maritime Safety and Security Towards
Economic Prosperity Conference"
Thursday, October 14, 2010,
Millennium Hotel
Stuttgart, Germany

Closing remarks by African Union Ambassador John K. Shinkaiye, 2010

CHAPTER 7

Post Retirement to Present

Event: Retirement Ceremony for General William "Kip" Ward
What: An unforgettable, soul-satisfying retirement celebration!
Location: Fort Myer Parade Field, Virginia
Date: April 28, 2011

Packing the grandstands in every direction I looked were people who came to celebrate me and my career. There was no doubt in my mind that just as Joyce was with me on the reviewing stand, they were there to celebrate her too, and rightly so.

The formation of soldiers on the parade field included a battalion from the 3rd Infantry, the Old Guard, Commander of the Old Guard, and the Army Band, "Pershing's Own." In the traditional uniforms doing the "duck" step was the Fife and Drum Corps and the Continental Color Guard. The Secretary of the Army was the ceremony host, with the Chief of Staff of the Army and other service chiefs, senior officers, and civilians there in attendance for my retirement celebration.

From the very beginning I knew this would be an amazing retirement celebration. My emotions swelled as Joyce and I climbed up the steps to the reviewing stand. I saw family and friends from early childhood to college, our military family, and so many others who had been with me throughout my 40-year career. This meant so much to my family who were present: beginning with Joyce, my mother, sister, son, daughter, granddaughter, in-laws and cousins. My Baltimore family even chartered a bus to come down to Fort Myer for this celebration.

Looking over to the grandstands, I saw teammates, fellow soldiers, officers, noncommissioned officers, and so many personnel I had served with over the years. Friends came in from

Hawai'i, New York, North Carolina, Georgia and even Germany to celebrate. As retirement ceremonies go, this was already a major success in my eyes.

On occasions like this, the honorees make comments and acknowledge colleagues and family who have impacted their career journey. I recognized them all: our children, my mother, other family members, military families, civilian colleagues, noncommissioned officers, soldiers, and those participating in the retirement review.

While I was at the podium, I took the opportunity to show that I am a product of the Motown Era. Now, I like to sing but I can't sing. In most cases that doesn't stop me. When I was in Command and General Staff College, our classmates would have our little social get-togethers. I'd grab a broom handle and start singing. It didn't matter that I couldn't hold a tune.

As I gave my remarks, I acknowledged the person who had been on this glorious journey with me the entire time, for every assignment and promotion, every deployment and missed holiday and special event. Joyce Lewis and I had gotten married just before our graduation from Morgan. She and my mother along with General Chappie James had pinned my lieutenant's "butter" bars on my lapels. I entered active duty with my wife and lifelong partner. Other than combat and hazardous duty tours, everywhere I went, she was with me.

In honoring Joyce Lewis Ward—my best friend and partner in life—I used my favorite song by The Temptations. I knew I couldn't get up on that podium in front of a thousand soldiers and 300 guests in the stands and start singing—*You're welcome!* Instead, I used a singing greeting card. As I spoke about my wife, I opened the greeting card, and it began playing *"My Girl"* over the microphone and PA system. People started laughing and clapping in the stands— the troops out on the field—everybody got a kick out of it, especially those who knew that it was my favorite song.

During that ceremony, I made a couple of non-standard

moves. For the final inspection, the trooping of the line of soldiers on parade, the Commander of Troops (who was the commander of the 3d Infantry Brigade - the Old Guard) came forward and invited me to conduct my final inspection before retiring. (It's the same unit seen at Arlington National Cemetery with the caissons, etc.) I marched out to "*Troop the Line,*" acknowledging the various formations along the way. As I'm walking along the line of troops with the Old Guard Commander of Troops, the 3rd Infantry Unit of the Old Guard, and in passing each unit on parade, I see the faces of soldiers, representing those I've served with over my 40-year career.

During that final inspection I, as the reviewing officer, walked along the line of troops, rendered a hand salute when passing the national colors, and acknowledged the other formations. I added another twist – as I walked past the battalion of infantry troops, I ducked three or four rows deep into each company formation, and shook a soldier's hand, thanking him for his service. I ducked in and out of the line of troops, shaking hands with soldiers along the frontage of this formation. That had never been done before in the middle of such a formal ceremony. I just wanted to shake the hands of soldiers—male and female—who had been my teammates, to personally touch those soldiers and tell them thank you.

The Old Guard Commander walking with me told me that was the first time he'd seen something like that, and his troops loved it. That retirement ceremony reflected my career: being with my soldiers, never taking them for granted and respecting them, and paying tribute to Joyce and our family.

A couple of months after the formal military retirement ceremony attended by the busload of family members and the host of military related colleagues and friends from all over the world, Joyce and I had another surprise. My sister Christina and first cousin Lawrence Gross (maternal cousin) gave us a wonderful retirement party in Baltimore. In addition to all of the Cashen family (maternal) in Baltimore, my Ward family (paternal) from multiple locations (Virginia to Costa Rica and California) were also there in force. I was happy to see all of them along with boyhood friends and college classmates, some of whom I hadn't seen in over 40 years. I think Phyllis, and my Aunt Hilda had more fun at the party than anyone. That Joyce, KJ and Kahni were there celebrating too was not lost on anyone. They had served and this was their party too. I still smile when thinking about it.

Satisfying Business Leadership

While I had retired from active duty, I was certainly not ready to retire from work. I had the privilege to serve as the president and COO of SENTEL Corporation, an engineering, IT, logistics and intelligence services firm headquartered in Alexandria, Virginia. They develop solutions for spectrum management and surveillance systems; chemical, biological, radiological, and nuclear detection systems. Over almost five rewarding years at SENTEL, I learned many things about the corporate world and business; especially gratifying was that SENTEL supported the Warfighter, a mission dear to me. I learned the business, contracts and procedures, and the rules associated with finance, payroll, HR, pipeline, business proposals, profit and loss—all elements that make a business work. Initially, my knowledge of how companies and businesses worked was not strong, but working with and learning from teammates added value to what was being done and made a difference. It's called *leadership.*

A former military teammate had learned what I was doing in the corporate world and that SENTEL did engineering work as well. When we met he said, "Sir, you're not an engineer. You're an infantryman. How do you know what you are doing in this environment?"

I said, "I know how to lead, and I know how important that work is; and I know the people that do it are important. I'm here to do my best to see that they are empowered to do their job as best they can, thus making all of us better."

We got a chuckle out of that but it's exactly right. His question reminded me of those assignments during my early military years. Important for me was—whatever the topic or procedure—making sure to learn enough about it so that my understanding was deep enough so that I wasn't disruptive to the process or the objective. My knowledge also needed to be deep enough to add value to the process, to provide guidance and direction that would enhance whatever was going on.

The SENTEL team seemed to appreciate that. I didn't walk around knowing it all. I would listen to what they would say, understand things from their point of view and then based on my ability to interpret, translate and then see it from different perspectives, relating it to other parts of the business and causing that part of the business to work better. That was critically important, making the whole better because the parts were working better together.

In leading organizations, whether civilian or military, success depends on how people perform and what they do. It is the job of the leader to ensure that persons on the team doing the work—be they soldiers, sailors, airmen or marines; be they technicians, contract specialists, engineering specialists, logisticians; be they air traffic controllers or information technology specialists—understand their roles and respect the roles of others. Those specialists who do their jobs day in and day out are truly the backbone of what makes a company go, just as soldiers, sailors, airmen, marines, and civilian personnel are the backbone of what makes a command go. Those factors are constants for how you treat and take care of people so that they know that those who are leading them have their interests at heart in all that they do. That part of becoming a corporate executive was an easy transition for me. There are parallels in what it means to take care of a person and doing it in a way that every person in every position knows that they are valued as individuals.

Essential to being an effective leader is ensuring that members of the team know they are valued individuals and know it in ways that matter; they know it's not just lip service. It needs to be demonstrated consistently, and that is one key to having high-performing and successful organizations. That was my leadership style, the team continued to perform with excellence. This was done for the success of the customers that we were supporting and to the advantage of the teammates on the team. SENTEL was doing well.

Given my military career that focused on sustaining peace, it was important to me to desire a global common, where conflict and instability are reduced. I looked forward to being connected with boards and groups that were connected to familiar values. Organizations like the Council on Foreign Relations, the Atlantic Council, and the Corporate Council on Africa (CCA), an entity that looks to promote American/African business and enterprise development on the continent of Africa.

As a career Army officer, my goal was protecting United States interests and security. As the commander of AFRICOM, I worked to bring opportunity (in service of achieving stability) to the African continent and its island nations. My post corporate life activities sustained that focus. My role as an advisory board member at CCA has been a way for me to continue serving this mission from a different vantage point. The same has been true with my other council and board affiliations.

Having a secure and stable environment is more than just strong militaries. Security supports how people feel in their homes and communities; it is the assurance of having better health conditions, the ability to earn a living, and to pursue education at any level. Those values are important to people around the world. From Europe to Asia to the Pacific to Africa and the Middle East, this is the case; and my assignments around the world showed me that people are more alike than they are different, regardless of their race, their religion, or cultural differences.

About That Horizon of Hope

A *horizon of hope* exists when people all around the world have a stable and secure environment, are well cared for and are free to fulfill their life's potential. Do people have a horizon of hope where they are? Is there a horizon of hope for them such that tomorrow they see something better than today, for them and especially for their children and grandchildren? Are economic development and social enhancements in education and health initiatives in place and maintained that will make a difference to people wherever they are?

Some organizations emphasize the role of development. Business and economic development is important in supporting meaningful and dignified ways for people to provide for themselves and their families. Nutrition is also critically important. I am grateful that in my post-corporate life, organizations with these purposes are growing. I continue consulting and advising corporations who work with and seek out business opportunities in various parts of the world. In a limited way, this keeps me relatively busy. It is not full time, but not a day goes by when I don't feel as if I might be able to do something to help someone else.

Life Today

Joyce and I decided on where we would settle down. Typically, that decision is based on the answer to one of two questions: Where do you want to live or what do you want to do? The answer to one would determine the other. We knew we liked the DMV (District of Columbia, Maryland and Virginia) area, notwithstanding the cost.

It would be either the location or the occupation—in our case it was the location. We decided we wanted to settle in the mid-Atlantic region. On the east coast we had family, friends and familiar environments; that area also served as a launch pad for doing things on either side of the Atlantic. We liked the Virginia side because it gave us access to Dulles Airport; I knew I might be connected with entities that would involve travel, especially international travel.

We chose the Northern Virginia area. It's funny. I kept hearing about this DMV thing, thinking it was about the Department of Motor Vehicles. After spending nine years overseas, I began to lose track of acronyms and abbreviations. It took a minute to connect the national capital region (NCR) with this DMV moniker.

Our post-military life has been filled with my stint in the corporate world as president and chief operations officer at SENTEL. I remain engaged in international affairs reflective of my senior general assignments through organizations such as the Council on Foreign Relations, the Atlantic Council, the Hudson Institute, the Center for Strategic and International Studies, the U.S. Global Leadership Coalition, and the Corporate Council on Africa. I also get to satisfy my interest in teaching through periodic lecturing at universities and other speaking engagements – reminiscent of my days at West Point.

Joyce continues her involvement with childhood education, an absolute passion of hers and something she is continually devoted to. As a military spouse, she paid a lot of attention to the experiences of military-connected families, and she remains committed to those efforts today through various organizations. From board work with the Military Child Education Coalition to a half dozen scholarship programs, she's administered and raised funds for several nonprofit organizations, continuing to improve her foxhole as well.

We are heavily engaged in our fraternal/social/volunteer organizations, and in paying it forward to our nation's next generation. My fraternal organizations have the same purpose at heart in being involved in ways that help uplift young people. And finally, we are blessed to be close to family and friends and enjoy being a part of their lives.

General William "Kip" Ward's Video Clips

Video Title	Description	QR Code
General Ward Reflects on his Army Service: TEDx Talk, Pentagon	General William "Kip" Ward, USA "Footprints: The Legacy We Leave"	
Leadership Series: General William Ward	General Ward's leadership style as commander of AFRICOM (2007-2011)	
Army Today Interview: General Ward	Gail McCabe interviews General Ward, commander of US Africa Command	
"From Battlefield to Boardroom" General William Kip Ward and Mr. Bernard Height	General Ward and Mr. Bernard Height explore the evolution of leadership from military service to roles in private sector.	
Center for Strategic & International Studies: Video Highlight	Military Strategy Forum: General William E. Ward, Commander, U.S. Africa Command	

Diversity in National Security Network	To elevate diverse voices, Gilberto Washington interviews General Ward	
Holiday Greetings from Commanding General William "Kip" Ward	General Ward and Mrs. Joyce Ward send season's greetings to troops around the world, from US AFRICOM	
Foreign Policy Research Institute Video Message	Ward's message to teammates in 2010, focusing on the Foreign Fighter phenomenon, with recent trends and case studies.	
IUEC Chronicling Greatness	Reflections on September 11, 2001	
US Army General Ward at The Trumpet Awards	Ward receives award highlighting African Americans who have achieved success in all walks of life	
The Challenges of AFRICOM	Former AFRICOM commander, General Ward, explains challenges of U.S. Army and AFRICOM.	

Black Engineer of the Year Award	General Ward honored with the Lifetime Achievement Award at the 24th Black Engineer of the Year Awards, Baltimore, Md., 2010	
Black History Month – US Army General Ward	US AFRICOM Commander, General Ward discusses the benefits of diversity in the Army.	
AFRICOM Priorities – News Report – 090417	General William Ward explains how U.S. forces work with partner nations in Africa.	
General Kip Ward: Outstanding Role Model	Ward remarks on how having confidence, belief and preparation, and standing on shoulders of ancestors is a tremendous gift.	
General Ward on Higher Education	Ward explains how the education gives a foundational support that opens doors, expands imaginations and boosts confidence.	
General Ward at Diversity Leadership Conference, Part 1	Ward encourages critical thought with persons who expand horizons, that achieving diversity is a never-ending process.	

General Ward at Diversity Leadership Conference, Part 2	The General explains that diversity is essential and critical to our moving forward.	
General Ward at Diversity Leadership Conference, Part 3	A description of how AFRICOM, as a unified command, has a role to play in the stability of the continent of Africa and its island nations.	
General Ward receives the Visionary Award	The Africa Center for Strategic Studies (ACSS) presents General Ward with the 2011 achievement award for peace and security.	
General William E. Ward AFRICOM Commander visits Algiers	General Ward and Algeria President Abdelaziz Boutefilka discuss issues and solutions for cooperation between Algeria and the U.S.	
Overcoming Adversity: Improving Your Foxhole with General Kip Ward	Ward's comments reflect on facing adversity and the important role of faith, friends, and family.	
General William Ward Press Conference in Algiers	From General Ward's trip to Algeria, working strengthening and expanding mutual cooperation interests.	

AFRICOM Now 19	Gathering of air chiefs from more than 20 African nations met with Army groups to establish friendships and relationships and improve medical personnel readiness.	
ACSS Visionary Award 2010: Speech – General William E. Ward	ACSS Visionary Award 2010: Speech - General William E. Ward	
ACSS Visionary Award 2011: Acceptance Speech—General William E. Ward	The visionary award highlights his contribution toward promoting democracy and improving civil military relations.	
ARISE News: Former US Secretary of State Colin Powell Dies – General William Ward & Eric Ham	General Ward and leaders honor the legacy and example of the late US Secretary of State Colin Powell; a soldier, a statesman, a public servant.	
Senator Gerald Neal (Kentucky Senate Democrats) on Senate Resolution General Charles Young	Retired Col. William "Kip" Ward and General David Jenkins honors Brigadier Army General Charles Young for courage and service.	

WPSC 2011 – General Ward Answers Questions from Delegates	At the 2011 West Point Security Conference. Ward responds to questions about U.S. military actions with humanitarian activities and possible solutions.	
US Southern Command, Northern Command, Africa Command, and Joint Forces Command (Part 1 of 2)	Representing AFRICOM, General Ward is among several commanders examining global conflicts and missions.	
General William Ward of AFRICOM Visits Algeria	General Ward explains the U.S.'s support for Algeria's counterterrorism efforts; listens to Algerian president's recommendations.	
General William Ward Interview on Algerian TV, Part 1	In Part 1, Gen. Ward explains the U.S.'s primary responsibility as it relates to security and its commitment of with African nations.	
General William Ward Interview on Algerian TV, Part 2	For Part 2, Gen. Ward discusses processes of stability and being partners with Algerian and other African communities.	
Frost Over the World – Jordan and Ward – Oct 24 – Part 2	2008 Program: General Ward discuss AFRICOM and how it coordinates American military and security interests, covering all of Africa, except Egypt.	

Gamma Pi Chapter of Omega Psi Phi Fraternity, Founder's Day Keynote Address	From 2012, General Ward's message honors the fraternity and celebrates unity during Founder's Day and fraternity's 101 years.	
Nigerians Celebrating top US Military General	General Ward is inducted into the Igbo nation in Nigeria.	
General Ward "For the Love of Liberty" Message	At Fort Leavenworth, Kan. in 2010, General Ward introduces the documentary film honoring African American soldiers throughout history.	
As The World Remembers D-Day #RMU Honors Black Soldiers Mainstream Media Failed to Mention	On "Roland Martin: Unfiltered," General Ward speaks on the sacrifices soldiers of color made during D-Day and throughout World War II.	
Goals of AFRICOM	General Ward talks with Gail McCabe about AFRICOM's commitment through sustained security and engagement to helping African nations build the capacity they want for stability and security.	

Overcoming Discrimination	Representing AFRICOM, General Ward talks about moving forward, overcoming hardships, and staying prepared.	
Leadership Series: Remembering Colin Powell	General William Ward Remembering Colin Powell	
Qureia meets new US Security Coordinator	Qureia meets new US security coordinator	
Foreign Affairs Hearing: US Interests in Africa	Members from the National Geographic Society and Johns Hopkins University question General Ward about global terrorism, particularly in Africa.	
Your Next Mission Season #3 EP 22—Celebrating Black History Month 2023, "Black Resistance"	CSM Tonya Oxendine and General Ward discuss how African Americans have resisted historic and ongoing oppression while making monumental contributions.	

ADDENDUM:

Timeline of Education and Military Service

Early Life and Education

June 1949 - June 1971

Early childhood years to graduation from Morgan State College as a distinguished military graduate with a degree in Political Science, Baltimore, Md.

Military Training and Initial Assignments

June 1971 - November 1974

Student, Infantry Officers Basic Course, Fort Benning, Ga. Platoon Leader, Company Executive Officer, Anti-tank Platoon Leader, Battalion Motor Officer, Brigade Training Officer, 3rd Battalion (Airborne), 325th Infantry, and 2nd Brigade, 82nd Airborne Division, Fort Bragg, N.C.
Rank: 2nd Lieutenant and 1st Lieutenant.

November 1974 - April 1976

Executive Officer, B Company, and later, Commander, C Company, 1st Battalion (Mechanized), 17th Infantry, 2nd Infantry Division, Republic of Korea.
Rank: 1st Lieutenant and Captain.

Advanced Training and Education

May 1976 - December 1976

Infantry Officers Career Course, Fort Benning, Ga.
Rank: Captain.

January 1977 - January 1979

Graduate School Student, Pennsylvania State University, State College, Pa.
Rank: Captain.

January 1979 - June 1982
Assistant Professor, Department of Social Sciences, USMA, West Point, N.Y.
Rank: Captain.

July 1982 - June 1983
Student, Command & General Staff College (CGSC), Fort Leavenworth, Kan., Class of '83.
Rank: Captain and Major.

July 1983 - August 1983
Student, Supply Management Officers Course, Fort Lee, Va.
Rank: Major.

Leadership Roles in the Field

August 1983 - June 1985
Brigade Logistics Officer, S4, 210th Field Artillery (Heavy) Brigade, VII Corps Artillery, Herzo Base, Germany.
Rank: Major.

July 1985 - July 1987
Community Executive Officer, US Military Community Activity, and later, Battalion Executive Officer, 1st Battalion (Mechanized), 7th Infantry, Aschaffenburg, Germany.
Rank: Major.

August 1987 - November 1988
Student, Logistics Executive Development Course, Fort Lee, VA, and later, Staff Officer, Department of the Army Staff (Deputy Chief of Staff, Logistics), The Pentagon.
Rank: Major.

November 1988 - June 1991
Commander, 5th Battalion (Manchu), 9th Infantry, and

later, Division Logistics Officer, G4, 6th Infantry Division (Light), Fort Wainwright, Alaska.
Rank: Lieutenant Colonel.

Senior Leadership and Command

July 1991 - June 1992
Student, U.S. Army War College, Carlisle Barracks, Pa.
Rank: Lieutenant Colonel.

June 1992 - June 1994
Brigade Commander, 2nd Brigade, 10th Mountain Division (Light), Fort Drum, N.Y., and Somalia.
Rank: Colonel.

June 1994 - August 1996
Executive Officer to Vice Chief of Staff, Department of the Army, The Pentagon; later, Deputy Director Operations, J3, Joint Staff, The Pentagon.
Rank: Colonel and Brigadier General.

August 1996 - January 1998
Deputy Commanding General, 82nd Airborne Division, Fort Bragg, N.C.
Rank: Brigadier General.

January 1998 - July 1999
Chief, Office of Military Cooperation, U.S. Embassy, Cairo, Egypt.
Rank: Major General.

August 1999 - December 2000
Commanding General, 25th Infantry Division (Light), Schofield Barracks, Hawai‘i.
Rank: Major General.

December 2000 - October 2002
Vice Director for Operations, J3, The Joint Staff, The Pentagon (during 9/11).
Rank: Major General.

November 2002 - November 2003
Commander, NATO Stabilization Force, Sarajevo, Bosnia-Herzegovina.
Rank: Lieutenant General.
November 2003 - February 2005
Deputy Commander, U.S. Army Europe (USAREUR), Heidelberg, Germany.
Rank: Lieutenant General.

Final Positions

February 2005 - December 2005
U.S. Security Coordinator, Israel-Palestinian Authority, Tel Aviv and Jerusalem (Gaza and West Bank) (Special duty assignment reporting to Secretary of State).
Rank: Lieutenant General.

January 2006 - April 2006
Deputy Commander, USAREUR, Heidelberg, Germany.
Rank: Lieutenant General.

May 2006 - October 2007
Deputy Commander, U.S. European Command, Stuttgart, Germany.
Rank: General (The Army's Fifth Black 4-star General)

October 2007 - March 2011
Inaugural Commander, U.S. Africa Command, Stuttgart, Germany.
Rank: General (The nation's first Black Combatant Commander).

ACKNOWLEDGEMENTS

This book is dedicated to everyone who have been on this 40-year journey with me and to those who want to know about it. My parents, Isiah Richard Ward, Jr. and Phyllis Cashen Ward taught me about life, respect, hard work, dedication, humility and selfless devotion to enhancing the wellbeing of others just by observing their behavior. My maternal and paternal grandparents—Rev. Joseph and Marie Cashen, Richard Ward, Sr. and Mary Parker Ward Flood—reinforced everything my parents showed me as a youngster. For my sister Christina and me, they were the first examples and role models when it came to living an honorable life. My childhood neighborhoods of Falls Road and Bare Hills were the ultimate testimony to the value of a "village raising a child." Neighbors and friends had a profound impact on my life. Those experiences and lessons are enduring.

From the start, everything I achieved or accomplished was possible due to the loving support, devotion and uncommon dedication of my extraordinarily exceptional wife, my college sweetheart, the love of my life and the mother of our beautiful children, Joyce Lewis Ward. She, our son KJ and daughter Kahni were steadfast supporters and constant sources of encouragement in all circumstances. They were my inspiration to do well, even when we were apart. When we were apart, I knew Joyce was handling things without fail, from raising our children to taking care of our household as she was able to intermittently pursue her career. I would be nothing without them. I love them dearly.

With Joyce I was blessed to gain another family, the Lewis and Hopson crew. To Joyce's mom, Mamie Hopson Lewis, I was a hero; and what a beautiful and giving person she was. Joyce's sister, Doretha L. McBeth and her husband Douglas, trusted me implicitly. Joyce's brother, Levi Lewis, Jr. and his wife Paula, have been constant champions. Next to Joyce, KJ and Kahni, Levi has attended more of my ceremonies than anyone. Walking her down the aisle at our wedding, I guess Levi was making sure things were lasting. (Joyce's father passed away when she was eight years old; however, we were able to learn about him after meeting Tom Wilkins, who provided a unique perspective on Levi Sr., based on their young adult friendship in Lawrenceville, Virginia. We really appreciated that, as did Kahni and KJ when they heard Tom speak about their grandfather.)

My journey would have taken a far different turn had it not been for the mentorship and education I received at Morgan State College. My military cadre and civilian instructors instilled in me the absolute necessity of knowing your job, always seeking self-improvement and constant learning to be the best at whatever you're doing. Colonel Mel Hurt, Captain Nick Lagatutta, and Sergeant First Class Broadus Mattison were consummate professionals and caring leaders. Professors Gill, Walker, Adair and Franklin were educators who inspired learning.

Every soldier and leader with whom I've ever served has an ownership stake in my success. Beginning with my first assignment in the 82nd Airborne Division, Captain Sam Boyd provided valuable lessons of officer leadership and Sergeant First Class P. Gordy and First Sergeant E.M. Hunter demonstrated true noncommissioned officer (NCO) leadership. That NCO leadership was epitomized to the highest degree by the two best NCOs I ever knew, my Company First Sergeant in Korea, Alfred Hill and my Division Command Sergeant Major when I commanded the 25th Infantry Division in Hawaiʻi, Mark Ripka. Mark later was Senior Enlisted Leader when I became the nation's first Black Combatant Commander as the inaugural Commander of the U.S. Africa Command.

Each of my Aides-de-Camp, Staff/Administrative/Executive Assistants, Executive Officers, Security Officers and Enlisted Aides became and remain members of our family: Dave Hodne, Major Luckett, Reggie Buie, Gerald Hunter, Brandon

Alexander, Chris Crane, Mr. Osman, Gary and Betty Gariepy, Rhett Boldenow, Steve Parker, Garry Bishop, Bill Bender, Tom Cloma, Tracy George, Lyrica George, Susan Patrick, Dyfierd Harris, Jabbar Bruce, Doug Wilderman, Rob Harvey, Cheryl Harris, and Steve Michael, with me spending more time with them collectively than I did with Joyce, KJ and Kahni, especially over the last 16 years of my career as a general officer.

I am deeply indebted to everyone who has encouraged me to write this memoir for the past 10 years. Most of those named above along with so many others have pushed me to tell my story including, Joyce and Cortez Dial, Dennis and Linda Via, Bruce Moody, Lisa and Greg Packer, Erv and Pat Pearson, Juanita and Bill Harvin, Bob and Monique Ferrell, Jim and Yolanda Mason, Wendy and Dwight Dean, Bill and Michelle Puttmann, Song Li and Jim Kelley, and Melba Uzzell along with many members of my fraternities: Omega Psi Phi, Pershing Rifles and Sigma Pi Phi and Joyce's sororities: Delta Sigma Theta, The Links, Incorporated and The Society. Several of my Command and General Staff College classmates from the famed Class of 1983 have been sources of constant inspiration, Chuck and Kate Ware, Jake and Myralon Simmons, Don and Yvonne Glover, Gordon and Rhee Burke, George Ingersoll, Herb and Paris Tillery, Wil and Elaine Giles, Phil and Carol Wilkerson, and other members of our special group who continue to meet each year.

There are a host of mentors, advocates, bosses and colleagues who invested in me and to whom I am forever grateful. Flag officers, ambassadors and other senior officers including: Johnnie Wilson, Arthur Gregg, John Tilelli, Colin Powell, Jim Jones, Sam Ebbesen, H. Tom Fields, Steve Arnold, Julius Johnson, Charles Hood, Edward Myer, Jim Monroe, Julius Becton, Harvey Williams, Matt Zimmerman, Ted Britton, Don Scott, Mike Mullen, Jim Wood, Linda Thomas-Greenfield, Andy Chambers, Leo Brooks, Sr, Jendayi Frazer, Waldo Freeman, Warren Mercer, Clarence Miller, Dan Kurtzer, Roscoe Robinson, Joe Reeder, Mary Carlin Yates and Daniel (Chappie)

James, all had an instrumental and richly appreciated impact on some aspect of my career.

I ultimately committed to this book at the urging and with the great assistance of Dr. Gerald Curry and Curry Brothers Publishing. This team of professionals showed me the route to make this project a reality. I am deeply grateful for the encouragement and guidance they provided. A special thank you to my son who brought his editing expertise to bear in helping me put words to paper.

Finally for our children and their spouses, KJ, Kahni, Randy, Kenyatta, and our grandchildren Kamila and Kaleb, I dedicate this book to you so that you have a lasting account of the life we lived in service to our country and the global commons. My hope is that it will inspire you to continue to serve others, always seek self-improvement, be thankful for the blessings bestowed on you and strive to make a positive difference in all you do, treating everyone with dignity and respect.

CURRY BROTHERS
PUBLISHING